Metal–DNA Chemistry

ACS SYMPOSIUM SERIES **402**

Metal–DNA Chemistry

Thomas D. Tullius, EDITOR
Johns Hopkins University

Developed from a symposium sponsored
by the Division of Inorganic Chemistry
of the American Chemical Society
at the Third Chemical Congress of North America
(195th National Meeting of the American Chemical Society),
Toronto, Ontario, Canada,
June 5–11, 1988

CIC
SQM
IMIQ
AFM
ACS

American Chemical Society, Washington, DC 1989

Library of Congress Cataloging-in-Publication Data

Metal–DNA chemistry
    (ACS Symposium Series, 0097–6156; 402).

    Developed from a symposium sponsored by the Division
of Inorganic Chemistry at the Third Chemical Congress of
North America (195th Meeting of the American Chemical
Society, Toronto, Ontario, Canada, June 5–11, 1988.

    Includes bibliographies and indexes.

    1. DNA—Congresses. 2. Organometallic compounds—
Congresses.

    I. Tullius, Thomas D., 1952–   . II. American
Chemical Society. Division of Inorganic Chemistry.
III. Chemical Congress of North America (3rd: 1988:
Toronto, Ont.) IV. American Chemical Society. Meeting
(195th: 1988: Toronto, Ont.) V. Series.

QP624.M48   1989   574fm.3282   89–15135
    ISBN 0–8412–1660–6

Copyright © 1989

American Chemical Society

PRINTED IN THE UNITED STATES OF AMERICA

QP624
M481
1989
CHEM

ACS Symposium Series

M. Joan Comstock, *Series Editor*

Foreword

The ACS SYMPOSIUM SERIES was founded in 1974 to provide a medium for publishing symposia quickly in book form. The format of the Series parallels that of the continuing ADVANCES IN CHEMISTRY SERIES except that, in order to save time, the papers are not typeset but are reproduced as they are submitted by the authors in camera-ready form. Papers are reviewed under the supervision of the Editors with the assistance of the Series Advisory Board and are selected to maintain the integrity of the symposia; however, verbatim reproductions of previously published papers are not accepted. Both reviews and reports of research are acceptable, because symposia may embrace both types of presentation.

Contents

Preface

NOT MANY YEARS AGO a book dealing with the chemistry of both metals and DNA would have seemed a strange collection indeed. Inorganic chemistry and molecular biology were not often subjects of study in the same academic department, let alone by one individual. But a number of developments in both fields have made the connection between transition metal and nucleic acid chemistry a vigorous and exciting area of modern interdisciplinary science. The aim of this book is to introduce inorganic chemists and molecular biologists to the leading edge of research in metal–DNA chemistry.

Although metal ions have long been recognized as critically important components of biological systems, emphasis in the past has been placed on metals in association with protein. Only recently has the rich chemistry of metals and nucleic acids begun to be appreciated and exploited. The advent of simple platinum complexes as powerful antitumor drugs undoubtedly sparked much of the interest in this field. More recently, two other aspects of metal–nucleic acid chemistry have attracted great attention and are poised for explosive growth: the development of metal complexes as tools for the structural and functional dissection of genetic systems, and the discovery of metalloproteins that are components of gene regulatory systems.

Most of the chapters in this book represent contributions from a symposium titled "Transition Metal–Nucleic Acid Chemistry," whose goal was to enlighten chemists and molecular biologists in new areas of the chemistry of metal ions and nucleic acids. Previous symposia at national meetings of the American Chemical Society have emphasized the use of metal complexes in medicine. Three chapters in the volume present recent results in the chemistry of platinum, palladium, and other metals with DNA and nucleotides.

Two emerging research areas that have not previously been the subjects of ACS symposia are also featured. The chemistry of transition metal complexes has become the key to development of sophisticated and sensitive tools for studying structural and functional details of genetic systems. Unusual DNA structures can be unraveled, and protein–DNA interactions can be mapped at high resolution, using metal chemistry. Among these new tools for molecular biology are iron(II) EDTA, copper phenanthroline, and various metalloporphyrins.

The recently recognized role of metals in gene regulation provides an exciting new direction for research in inorganic chemistry and molecular biology. Mercury metalloregulation, proteins that bind to DNA and RNA using zinc fingers, and the iron-dependent Fur regulon are three systems that represent the advancing front of this field of research.

Besides presentations of new experimental results, the symposium also included tutorial sessions that served to introduce methods and ideas from both inorganic chemistry and molecular biology to specialists in the other discipline. Material presented in the tutorial is represented in several of the chapters of *Metal–DNA Chemistry*.

For financial support of the symposium, I gratefully acknowledge the Donors of The Petroleum Research Fund, administered by the American Chemical Society; the Division of Inorganic Chemistry of the ACS; E. I. du Pont de Nemours and Company; Engelhard Industries; and Procter & Gamble.

THOMAS D. TULLIUS
Johns Hopkins University
Baltimore, MD 21218

May 21, 1989

Chapter 1

Metals and Molecular Biology

Thomas D. Tullius

Department of Chemistry, Johns Hopkins University, Baltimore, MD 21218

The use of transition metal complexes as tools for molecular biology is reviewed. Two basic strategies are detailed: the design and synthesis of metal complexes that recognize, bind to, and cleave DNA, and the use of a metal complex [iron(II) EDTA] that cleaves DNA non-specifically through generation of free hydroxyl radical. Application of the latter strategy to the determination of the helical periodicity of viral promoter DNA, and structural details of bent DNA from a trypanosomatid parasite, are presented.

Understanding how metal complexes interact with DNA has become a central question in some of the most active research areas at the interface between chemistry and molecular biology. Three kinds of metal-DNA systems currently are of particular interest: metal complexes that are used as tools for molecular biology, metalloproteins that regulate gene expression by binding to DNA, and metal complexes that act as drugs. These systems span a wide range in complexity, from simple transition metal chelate complexes, known to the inorganic chemist for many years (such as iron(II) EDTA and the bis- and tris(phenanthroline) complexes), to more elaborate complexes specifically designed to bind to DNA [such as methidiumpropylEDTA•Fe(II)], to metals coordinated within proteins. All of these systems have the common feature that the chemistry of the metal is essential to the ability of the complex to interact with DNA.

Since the work of my own research group has been centered on using a simple metal complex, iron(II) EDTA, to derive detailed information on the structure of DNA (1), in this chapter I will concentrate on the use of metal complexes as new tools for the molecular biologist to use to dissect genetic systems. Some of these metal complexes are also the subjects of other chapters in this volume, so my description of them here will be limited. Since metal complexes that are used as drugs, particularly *cis-*

0097–6156/89/0402–0001$06.75/0

diamminedichloroplatinum(II) (Cisplatin), are familiar to most readers, and have been featured in other volumes in the ACS Symposium Series (2), I will forgo discussion of them. The chapters in this volume by Marzilli et al. and by Sigel present new results on the modes of platinum binding to DNA. As well, metalloproteins that are involved in gene expression are of wide current interest, and are covered in the chapters by Berg, Neilands, and O'Halloran.

Metal Complexes as Tools for Molecular Biology

Structure in Genetic Systems. Over the past thirty years we have progressed in our understanding of genetic systems from the theoretical prediction of gene repressors by Jacob and Monod (3), to the discovery, isolation, and sometimes even crystallographic characterization of a host of DNA-binding proteins that regulate gene expression (4). In parallel, appreciation of the structural subtleties of DNA has advanced from the simple picture of the double helix of Watson and Crick (5), to the discovery and structural characterization of of a variety of unusual DNA structures including left-handed Z-DNA, supercoiled circles of DNA, bent DNA, and three- and four-stranded DNA molecules (6).

Molecular biologists are used to considering DNA as a sequence of letters that represent the four nucleic bases. New protein binding sites (often called 'boxes') are located by searching for similarities in nucleotide sequence to known binding sites. Implicit in these studies is a neglect of the underlying three-dimensional structure of the DNA. It is now becoming clear that, far from being a passive, uniform substrate to which proteins bind, DNA contributes to the specificity of protein interactions by its ability to exist in alternate conformations (7), or to deform its structure to accomodate protein binding (8, 9, 10). This conceptual advance has arisen from detailed structural studies of DNA and DNA-protein complexes. Further progress in working out the intimate mechanics of gene expression and development will come from more general incorporation of such ideas into molecular biology.

Structural Tools for Genetic Systems. What tools are available to the molecular biologist to investigate questions of structure in DNA systems? The standard structural methods of the chemist, X-ray crystallography and NMR, are being increasingly applied (10–14). One cannot overestimate the importance of a crystal structure of a DNA-protein complex in advancing our understanding of how proteins recognize specific DNA binding sites. A limitation of this experimental approach is the difficulty of determining a crystal structure for every DNA system of interest. A related problem is that functional genetic systems are complex, often consisting of large segments of DNA bound by more than one protein (15). Interactions of DNA sequences (presumably mediated by bound protein) over distances of more than a thousand base pairs are commonly observed in eukaryotic gene regulatory systems (16). These very large and complicated assemblies of macromolecules will be formidable experimental subjects for the crystallographer or NMR spectroscopist.

Because of the difficulty of applying standard structural methods to genetic systems, other experimental approaches have been necessary. DNA turns out to be especially amenable to chemical and enzymatic probing of its structure. The principle behind these methods is that a small molecule or enzyme is able to recognize structural features of DNA, and to report on this recognition by cleaving the DNA strand where the structure occurs (Figure 1). The structures sensed can be regions of single-stranded DNA, left-handed helical segments, cruciforms, and so forth (6). By this same strategy, proteins or small molecules bound to DNA can be mapped by noting where in the DNA molecule the cleavage reaction is inhibited; this method is called 'footprinting' (17).

The myriad techniques developed by molecular biologists for manipulating DNA (18) (cloning, restriction endonuclease digestion, gel electrophoresis, DNA sequencing) are used to prepare a system for these experiments. Any DNA sequence can be generated, either by cloning or by direct synthesis. High sensitivity and resolution are achieved by application of two related techniques, radioactive labeling and DNA sequencing (19).

Very small amounts of DNA (micromoles to femtomoles) can be studied by labeling one end of a DNA molecule with radioactive phosphorus (Figure 1) (18, 19). Besides allowing the experimenter to keep track of these tiny quantities of DNA, the radioactive label more importantly gives a benchmark from which to measure the sizes of the fragments that result from cleaving the DNA molecule. The particular nucleotides in the DNA sequence at which strand cleavage occurs are found by using gel electrophoresis to measure the lengths of the radioactive fragments produced in the cleavage reaction (Figure 1). These gels, the same as are used for DNA sequencing, are capable of resolving bands representing cleavages at every nucleotide within a DNA molecule several hundred nucleotides in length. Cleavage kinetics can be measured simultaneously at a large number of sites on the DNA molecule by analysis of the intensities of the bands on the gel.

Given the general methodology of using chemical or enzymatic reagents to map DNA structure and proteins bound to DNA, the challenge comes in choosing a particular reagent to answer the question of interest. For example, enzymes that are known to cleave single-stranded DNA are used to find the unpaired bases at the tips of hairpins (20).

An alternative to relying on natural enzymes that have specificity for particular DNA structures is to chemically synthesize a new probe of structure, tailored to provide the desired structural information. Chemists have become adept at designing reagents that are complementary to the various structures exhibited by DNA. Most often metal complexes are used in these experiments (21), because the geometry of a metal complex and the structure of the ligand can be readily manipulated to build in

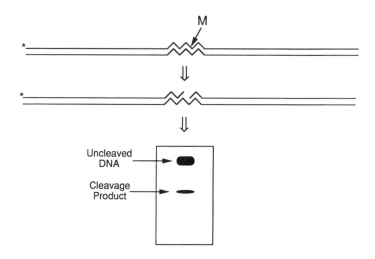

Figure 1. Chemical probes of DNA structure. The general strategy for such experiments is illustrated. A DNA molecule that contains an unusual structural feature (left-handed Z-DNA, single-stranded DNA, bent DNA, etc.), denoted by the wavy lines in the center of the DNA duplex, is radioactively labeled at one end of one strand (*) (top). A reagent M (chemical or enzymatic) that recognizes the unusual structure is used to cleave the DNA strand (center). The positions of cleavage (and thus the sites where the unusual structure occurs) are mapped by electrophoresis of the radioactively labeled, cleaved DNA molecules on a denaturing polyacrylamide gel (bottom). The lengths of the cleavage products can be compared to the lengths of standards to precisely identify the nucleotide(s) that adopt the unusual structure.

recognition features, and because the metal offers reactivity for cleaving DNA. Other strategies have been used, including the one developed in my laboratory in which the key feature is the *non-specific* cleavage of DNA by a metal-based reagent (1).

Tools for Manipulating Genetic Systems. Another important application of metal complexes to molecular biology is their use as replacements for enzymes that are used to manipulate DNA (22). For example, the restriction endonucleases have been of immense importance to molecular biology. Without these enzymes, the "cutting and pasting" of DNA sequences that is essential to recombinant DNA technology would be impossible. Natural restriction enzymes are limited, though, in the specificity with which they recognize DNA sequence. Four or six base pairs are the usual sizes of the recognition sites for these enzymes. While sites of this size are long enough to permit cutting of plasmid and viral DNA into a useful number of pieces, an entire genome is another level of complexity altogether. Mapping and sequencing the human genome will require reagents that are capable of recognizing sites much longer than six base pairs, so that a manageable number of fragments are produced in an initial digest. These smaller DNA fragments (but still large by viral or plasmid standards) then can be manipulated with other, less specific, reagents. While it is unlikely that natural restriction enzymes will be found that recognize 15 base pair-long sites, chemists have begun to engineer such specificity into synthetic, metal-based cleavage reagents (23–25).

Metal Complexes That Bind to DNA

The shape and the chemical structure of DNA provide a number of opportunities for interaction with metal complexes. The negative charges of the phosphates that are regularly spaced along the DNA backbone mediate electrostatic interaction with metal complexes. DNA has two grooves, major and minor, in which covalent, hydrophobic, and hydrogen bonding interactions can occur. The DNA base pairs, stacked perpendicular to the axis of the double helix, offer sites of intercalation for flat aromatic groups. Even the chiral nature of the DNA helix (whether it is right- or left-handed) has been used to direct selective binding of chiral metal complexes.

Copper Phenanthroline. The first 'artificial nuclease,' or metal complex capable of cleaving DNA, was bis(1,10-phenanthroline)copper(I), discovered by Sigman and his coworkers (26). The use of this complex in studies of DNA structure is described in detail in a later chapter in this volume. Copper phenanthroline binds in the minor groove of DNA, and then cleaves the DNA backbone by copper-mediated oxygen radical chemistry (27). This complex has been of particular importance in investigating prokaryotic transcription systems (28, 29).

MPE. A complex that was designed to first bind to and then degrade DNA is methidiumpropylEDTA•Fe(II) [MPE•Fe(II)] (Figure 2) (30). A molecule known to intercalate into DNA, methidium, was tethered to a metal-binding group, EDTA. This design permitted the delivery of oxygen radical chemistry to the DNA backbone (31). In this reaction the hydroxyl radical (•OH) is produced by reduction of dioxygen by iron(II) chelated by the EDTA group of MPE. The hydroxyl radical cleaves DNA by abstracting a hydrogen atom from a deoxyribose in the DNA backbone. The subsequent breakdown of the deoxyribose-centered radical results in the loss of the sugar and its attached base from the DNA backbone, causing the production of a gap in the DNA strand at the position of hydroxyl radical attack. The iron(III) product of the reaction is reduced back to iron(II) by added thiol, making a catalytic cycle (31). The combination of catalysis and propinquity allows the use of micromolar to sub-micromolar concentrations of MPE to efficiently cleave DNA (30, 31).

MPE•Fe(II) has been most widely used used to make footprints of small drug molecules and proteins that are bound to DNA (22, 32–35). Since binding of MPE•Fe(II) to DNA by intercalation precedes DNA cleavage, other molecules that are bound to DNA interfere with MPE•Fe(II) binding, and thus inhibit DNA cleavage. Because the hydroxyl radical is very non-selective in cleaving DNA, every nucleotide in a naked DNA molecule is cleaved by MPE•Fe(II). The MPE•Fe(II) cleavage pattern therefore gives information on the presence or absence of bound molecules for every nucleotide in the DNA. MPE•Fe(II) has the further advantage over the classical footprinting reagent deoxyribonuclease I that it is much smaller than the enzyme and so can more precisely delineate the edges of the footprint (35, 36).

Affinity Cleavage Reagents. MPE•Fe(II) is a non-specific reagent for DNA cleavage. Every nucleotide is cleaved at nearly the same rate, because the cleavage agent (the hydroxyl radical) is non-selective, and the DNA binding moiety (methidium) has no sequence preference. Dervan and coworkers have extended the approach pioneered by their work with MPE into a new realm, the attempt to impart sequence specificity to a synthetic DNA cleavage agent. They term this general approach affinity cleavage.

Their first experiments were in some ways the converse of previous work using MPE•Fe(II) footprinting to determine the DNA binding sites of the peptide antibiotics netropsin and distamycin. These drugs contain two and three N-methylpyrrolecarboxamide groups, respectively, that mediate binding in the minor groove of DNA. Footprinting experiments showed that the drugs preferred A/T rich sites. The lengths of the sites found led to a detailed model for recognition. Turning this experiment around, Dervan and coworkers then showed that a metal-binding moiety, EDTA, could be attached to the A/T-specific distamycin to give a reagent capable of much more specific cleavage than MPE, distamycin-EDTA

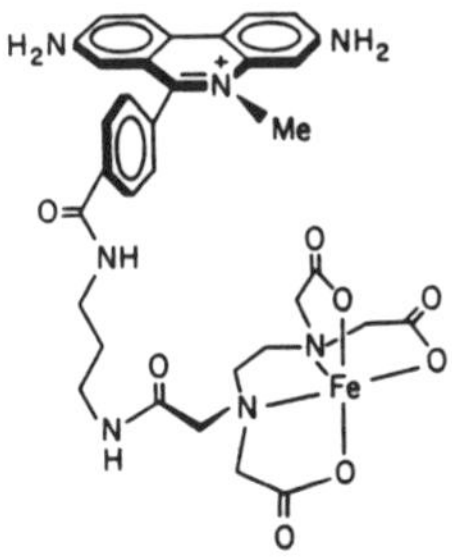

MPE•Fe(II)

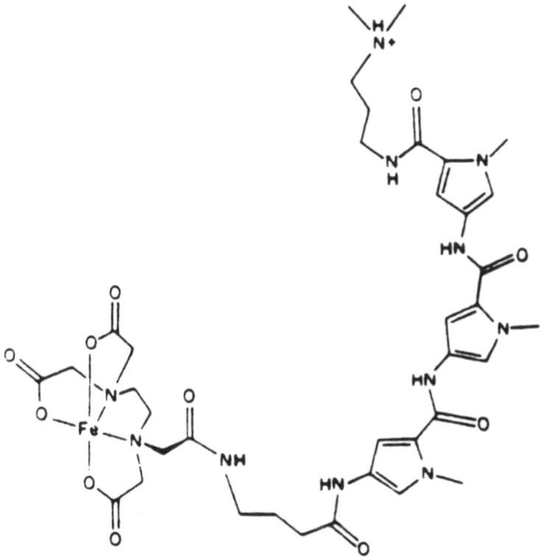

DE•Fe(II)

Figure 2. Metal complexes used to investigate DNA structure. Illustrated are methidiumpropylEDTA•Fe(II) [MPE•Fe(II)] (22), distamycin-EDTA•Fe(II) [DE•Fe(II)] (22). (Continued on next page.)

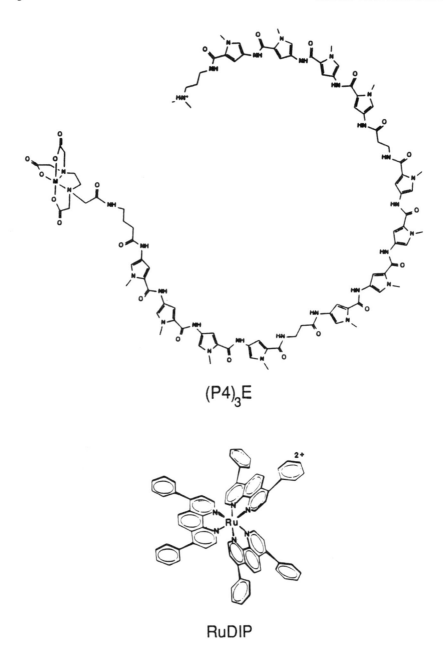

(P4)₃E

RuDIP

Figure 2. (Continued) Metal complexes used to investigate DNA structure. Illustrated are tri(tetra-N-methylpyrrolecarboxamide)bis(β-alaninyl)-EDTA [(P4)₃E] (23), and tris(4,7-diphenylphenanthroline)ruthenium(II) [Ru(DIP)] (21).

(Figure 2). Only cuts at A/T sites were observed, the same sites to which the free drug was found to bind (37).

With distamycin alone these sites were five nucleotides long. Homologating the DNA binding portion of the molecule led to an increase in the number of consecutive A/T base pairs necessary for binding and cleavage (38). With two distamycin units linked to EDTA, an eight base pair sequence 5'-TTTTTATA-3' was recognized (39). The crescent-shaped N-methylpyrrolecarboxamide moiety was found to be well-suited to act as a building block for even larger DNA-binding molecules that could take on a helical trajectory that would match the DNA helix. A molecule with EDTA connected to three tetra-N-methyl-pyrrolecarboxamide units, (P4)$_3$E (Figure 2), was found to bind, and cleave adjacent to, a site containing 16 consecutive A/T base pairs (23).

The ultimate way to achieve sequence-specificity is to use the hydrogen bonding capabilities of the DNA bases themselves to direct binding. In a number of different laboratories, DNA-cleaving moieties were attached to oligonucleotides, and then these modified DNA strands were hybridized to single-stranded DNA (40–47). Very specific cleavage at one or two nucleotides was found for copper phenanthroline attached to an oligonucleotide (41). An iron EDTA-modified oligonucleotide also gave specific cutting at the site of hybridization, but the cutting pattern formed a Gaussian-like distribution over four to six nucleotides (40). The difference in the copper- and iron-mediated cutting patterns is likely the result of different detailed mechanisms for cleavage. The iron system generates diffusible hydroxyl radical, while the copper cleavage chemistry is based on a copper-bound oxygen species that does not diffuse.

The strategy just described has the disadvantage that a single-stranded region in the target DNA is necessary for binding by the oligonucleotide-based cutting reagent. Since one of the most important uses for highly sequence-specific synthetic DNA cleavage reagents will be in making widely-spaced cuts in genomic DNA for mapping and sequencing the human genome, a preferable approach would involve a reagent that could cleave both strands of double-stranded DNA. It was recently demonstrated that the propensity for poly(purine)•poly(pyrimidine) sequences to form triple strands could be used to advantage in directing a sequence-specific metal-based cleavage reagent. Moser and Dervan found that a 15 nucleotide-long oligo(pyrimidine) strand, with attached EDTA group, was able to recognize and cleave a specific 15-base pair long oligo(purine)•oligo(pyrimidine) sequence within a 628 base pair long restriction fragment of DNA (24).

An important question that needs to be addressed for practical use of such reagents in genome mapping is the absolute specificity of the binding and cleavage. In other words, how much variation in sequence can be tolerated by the reagent and still result in cleavage? In a real genome, many sequences resembling the target sequence will likely be present. Strobel et al. recently showed using the triple-helix strategy (25) that an 18 nucleotide-long oligonucleotide-EDTA•Fe probe could recognize and cleave both DNA strands at a single site in the bacteriophage λ genome, which consists of 48,000 base pairs. The reaction could be done in an agarose gel, a useful property for studying very large genomic DNA.

Another way to build in high sequence specificity is to attach a cleaving group to a protein that binds to a specific DNA sequence. Sluka et al. attached EDTA to a 52-residue peptide that constitutes the DNA binding domain of Hin recombinase (48). The enzyme binds as a dimer to a specific 26 base pair-long site. The EDTA derivative of the Hin peptide cleaved specifically at the known Hin binding sites. In a related experiment, the tripeptide GlyGlyHis, known to be capable of binding copper, was attached to the amino terminus of the Hin peptide (49). Addition of copper to the hybrid protein resulted in a sequence-specific DNA cleaving reagent.

Chen and Sigman attached phenanthroline to the trp repressor and achieved sequence-specific cleavage upon addition of copper (50). This experiment implies that a single phenanthroline is sufficient to bind copper in a way that allows activation of oxygen by the metal.

Chiral Recognition of DNA by Metal Complexes. With the discovery of left-handed Z-DNA, the chiral nature of DNA came into sharp focus. Barton and coworkers (21, 51) recognized that the tris(phenanthroline) complexes of transition metals, long known to the coordination chemist, possessed chirality that might be complementary to that of DNA. Their work built on earlier studies of Lippard and coworkers who showed that terpyridyl complexes of platinum could act as metallointercalators, with the heterocyclic aromatic ligand intercalating between DNA base pairs (52). A similar mode of binding was found by Barton and coworkers for tris(o-phenanthroline)zinc(II) (53). This complex showed the additional feature that its optical isomers could discriminate in binding between right- and left-handed DNA molecules. Dialysis of right-handed B-DNA against the racemic metal complex resulted in enrichment of the D isomer inside the dialysis tubing, demonstrating selective binding of this isomer to right-handed DNA.

Greater chiral discrimination was found for complexes made with phenyl-derivatized phenanthroline ligands, and by using kinetically-inert metals like ruthenium (RuDIP, Figure 2) (54, 55). Cleavage chemistry was introduced into these systems by use of cobalt (56) as the metal: cobalt is known to cleave DNA by a photochemical reaction. Methylated phenanthroline ligands appear to provide selectivity for binding to the A form of DNA (57).

Other Metal Complexes that Bind to DNA. A remarkable observation is that tetraphenylmetalloporphyrins can bind by intercalation to DNA. Other metalloporphyrins are found to bind in the grooves of DNA. These results open the way for application of the multifarious chemistry of the porphyrins to DNA systems. Metalloporphyrin-DNA interactions are covered extensively in this volume in the chapters by Gibbs and Pasternack, and by Raner, Goodisman and Dabrowiak.

The Hydroxyl Radical as a Probe of DNA Structure

The common feature of all of the metal complexes so far described is that they first bind to DNA, before performing chemistry on DNA. My laboratory has pursued nearly the opposite strategy in our work on determining structural details of DNA systems. We decided to use the least-perturbing, smallest cleavage reagent possible to probe the structure of DNA by chemistry.

The cleavage reagent we chose is the hydroxyl radical, which was introduced above in the discussion of MPE and other iron-based DNA cleavage reagents. These complexes all deliver the hydroxyl radical to the DNA backbone after binding by intercalation or to the minor groove of DNA. For structural studies, a concern is that these modes of binding might themselves affect DNA structure. For example, intercalation of an ethidium molecule is known to unwind DNA by 26° (58).

We wanted instead to use the hydroxyl radical to investigate DNA in its native structural state. To do this, we also take advantage of metal chemistry, but in a way different than others have. A diagram of our experimental approach is shown in Figure 3. We use the negatively-charged EDTA complex of iron(II) to react with hydrogen peroxide in the Fenton reaction to generate the hydroxyl radical in solution. This metal complex exploits the negative charge of the DNA backbone, not for binding, but for electrostatic repulsion. $[Fe(EDTA)]^{2-}$ will not bind directly to DNA, but will instead sit in solution away from the DNA, reacting with hydrogen peroxide and bombarding DNA with the hydroxyl radical. The probe of the structure of the DNA is thus not the metal complex, but instead the hydroxyl radical molecule itself. The result of this experimental strategy is that the hydroxyl radical, a very small but also powerfully reactive molecule, gives a picture of the solvent accessibility of the surface of DNA.

As two examples of our approach, I discuss below experiments we have done to measure the helical periodicity of the promoter sequence of a viral gene (59), and to determine the structural basis for DNA bending (60, 61).

Chemistry and Structure Determination. Previous work by others had shown that the hydroxyl radical is capable of breaking the sugar-phosphate backbone of DNA. How often each nucleotide in a DNA molecule is cleaved may be measured easily by methods worked out for determining the sequence of DNA (19).

Further, the strand cleavage reaction mediated by the hydroxyl radical appeared to be the least specific of any reaction (enzymatic or chemical) that cleaves the DNA chain (52). Two characteristics of the hydroxyl

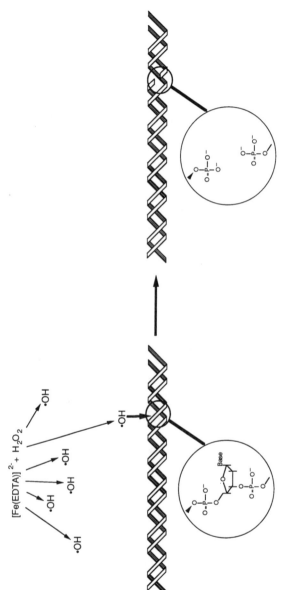

Figure 3. Strategy for the use of the hydroxyl radical for structural studies of DNA. The EDTA complex of iron(II) [Fe(EDTA)²⁻] reacts with hydrogen peroxide to generate the hydroxyl radical (•OH). Because both the DNA and the iron complex are anions, the metal complex does not bind to the DNA molecule, and thus the ultimate probe of structure is the hydroxyl radical. The hydroxyl radical cleaves the DNA backbone by abstraction of a hydrogen atom from a deoxyribose residue, resulting in the loss of the base and sugar at that position in the DNA strand. A single-stranded gap is therefore introduced into the DNA molecule. The phosphates remain that were originally connected to the deoxyribose that was attacked.

radical contribute to this lack of specificity. First, for the direct cleavage reaction, the site of attack on the nucleotide is the deoxyribose and not the heterocyclic base. Little if any selectivity for particular nucleotides would be expected. Second, the hydroxyl radical is known to react with organic substrates at rates that approach the diffusion limit (62), so cleavage of DNA should not depend on equilibrium binding of the reagent.

How can the hydroxyl radical be generated in the presence of DNA? Dervan's experiments had used iron(II) attached to other molecules to cleave DNA by a mechanism that was presumed to involve the intermediacy of the hydroxyl radical (31). Generation of the hydroxyl radical by iron(II) is known as the Fenton reaction, discovered at the end of the nineteenth century (63). In this reaction, iron(II) reduces hydrogen peroxide by one electron to give hydroxide ion and the hydroxyl radical (Equation 1):

$$Fe(II) + H_2O_2 \rightarrow Fe(III) + OH^- + \bullet OH \qquad (1)$$

Many complexes of ferrous ion can perform this reaction. Complexes of neutral ligands with ferrous or ferric ion will be cationic, of course, and would be expected to bind electrostatically to the polyanionic DNA molecule. In order to ensure that the only species in the system that can interact with DNA is the hydroxyl radical, we use the negatively-charged iron(II) EDTA complex in a version of the Fenton reaction developed by Udenfriend (64) (Equation 2):

$$[Fe(EDTA)]^{2-} + H_2O_2 \underset{ascorbate}{\rightleftarrows} [Fe(EDTA)]^{1-} + \bullet OH + OH^- \qquad (2)$$

The anionic iron chelates will not bind directly to DNA, so the hydroxyl radical must diffuse from some distance to the DNA before cleavage can take place. A further feature of the Udenfriend system is the use of ascorbate ion to reduce the iron(III) product back to iron(II), so that only micromolar concentrations of iron complex are required to cleave DNA.

Mechanism of DNA Backbone Cleavage by the Hydroxyl Radical. Although the hydroxyl radical undoubtedly attacks the heterocyclic bases as well as the deoxyribose residues in a DNA chain, the strand breaks that we detect are caused only by reaction with the sugars. The hydroxyl radical initiates chain cleavage by abstraction of a hydrogen atom from a deoxyribose. The deoxyribose radical intermediate then decomposes to give as final products two DNA chains terminated by the 5' and 3' phosphate groups that originally were adjacent to the deoxyribose that was the site of hydroxyl radical attack (Figure 3). We have determined that the DNA strands have phosphates at their ends by analysis of the mobility on a denaturing polyacrylamide electrophoresis gel of the radioactively end-labeled products. More work is necessary in order to describe in detail the mechanism of cleavage of DNA by the iron(II) EDTA/hydrogen peroxide/ascorbate reagent, but these efforts will be rewarded by increased insight into the structural details of DNA revealed by the frequency of cleavage at a particular nucleotide.

Studies of DNA Structure

Helical periodicity. The first image that comes to mind when you think of DNA undoubtedly is the double helix. The DNA helix is a central icon of modern biology, because this simple chemical structure embodies the essence of genetics. A characteristic of a helix made up of repeating units is the number of those units per helical turn. Different families of DNA conformation have particular helical periodicities. We became interested in developing a way to measure the number of base pairs per turn along any DNA molecule of interest.

An elegant experiment by Rhodes and Klug (65) gave us our starting point. They, in turn, had built on the observation by Liu and Wang (66) that DNA molecules uniform in length but random in sequence would bind to microcrystals of calcium phosphate, and that digestion of such bound DNA by an enzyme gave a modulated cleavage pattern. In the original experiment (66) these patterns were smeared out, because the electrophoresis gels used to separate the digestion products were not capable of resolving random sequence DNA molecules. Rhodes and Klug (65) used a gel formulation developed by Lutter (67) that gave single nucleotide resolution for such samples, and produced sinusoidal cutting patterns for random sequence DNA (extracted from nucleosome core particles) bound to calcium phosphate. They pointed out that the period of the cutting pattern corresponded directly to the helical period of the DNA, since the sites most exposed from the calcium phosphate surface (and thus most often cut) occur once each turn of the helix. Counting the number of nucleotides between each strong cutting site gives the number of base pairs per turn of the DNA molecule.

Because these experiments were done on random-sequence DNA, the helical period measured, 10.6 base pairs per turn, was the average for DNA in solution. We wished to map the helical periodicity along a DNA molecule of particular sequence. DNA molecules we hoped to study contained promoter and enhancer sequences involved in the regulation of transcription of a gene.

Although the general experimental approach seemed promising, there was a serious difficulty. In the Rhodes and Klug experiment an enzyme (usually deoxyribonuclease I) is used to cut the calcium phosphate-bound DNA. Unfortunately, deoxyribonuclease I cuts more often at some sequences than at others (68). If the DNA sample consists of a random collection of DNA molecules each with a different sequence, the enzyme will cut equally at each position since every possible sequence is represented at each position. But if we want to measure the helical periodicity of a DNA molecule of particular sequence, the cutting preferences of the enzyme will be superimposed on the modulation pattern caused by binding to calcium phosphate. This will lead to a very complex (and probably uninterpretable) pattern.

Our first application of hydroxyl radical chemistry to DNA structure determination (59) was inspired by the lack of cutting specificity that others had noted in iron(II)-mediated cleavage of DNA (31, 52). We reasoned that, if hydroxyl radical cleavage of DNA could be modulated by binding the DNA molecule to a calcium phosphate surface, then the resulting pattern would be a picture of the helical backbone of DNA. For our initial experiments we prepared a 200 base pair-long DNA molecule that consisted of the region upstream of the Herpes Simplex Virus-1 thymidine kinase (tk) gene. We were interested in the structure of this DNA molecule because several binding sites for proteins that regulate the transcription of the tk gene had been found in this sequence (69, 70).

We found that indeed a simple sinusoidal cutting pattern could be produced by cleaving the tk DNA molecule while it was bound to a calcium phosphate precipitate (Figure 4). The observation of a sine wave with a period of around 10 base pairs implied that the DNA molecule adopted one azimuthal orientation on the crystal surface. Rhodes and Klug had seen the same thing for random sequence DNA. They attributed the orientation specificity to the effect of unpaired nucleotides at the ends of the DNA molecule (65), which must therefore have a preferred way of binding to the calcium phosphate surface.

Our experiment gives a two-dimensional projection of a helical DNA molecule. What can we learn about DNA structure from this picture? The helical periodicity is the most obvious property we can derive. We found that the helix of the tk DNA molecule traversed 14 turns in 147 nucleotides, for an overall period of 10.5 base pairs per turn, characteristic of the B form of DNA.

We also mapped the positions of protein binding sites on the surface of the tk DNA molecule (Figure 4). This can help in understanding the protein-protein interactions that are necessary for control of gene expression. When several proteins participate in the control of one gene (as is the case for tk) it is hypothesized that these proteins must interact with each other, as well as with DNA, in order to have maximal effect on transcription (71). Since such proteins often are thought to bind to one side of the DNA helix, the interaction between two proteins is greatest if both lie on the same side of the DNA. The two-dimensional projection of the shape of the DNA helix that results from our experiment shows where on the DNA surface each protein binding site lies. The effects on transcription efficiency of deletion and insertion of DNA between protein binding sites may then be related to the relative positions of the regulatory proteins on the DNA helix.

Bent DNA. As part of the picture we have in our minds of the DNA double helix, DNA is visualized as a straight, rod-like molecule. Measurements in solution of the persistence length of DNA support this idea.

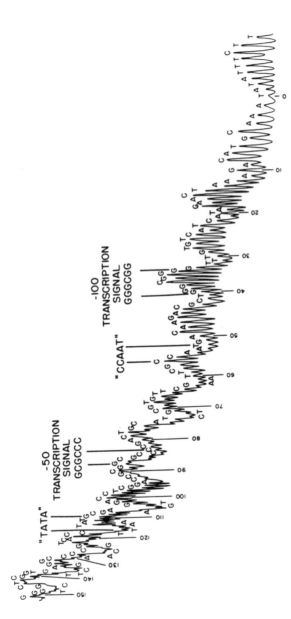

Figure 4. Two-dimensional projection of DNA helical structure via hydroxyl radical cleavage of DNA bound to a calcium phosphate precipitate. A 200 base pair-long singly-end-labeled restriction fragment of DNA from the Herpes Simplex Virus-1 thymidine kinase gene promoter region was used for this experiment. Cleavage of this DNA molecule while bound to a precipitate of calcium phosphate resulted in a modulated cleavage pattern, which has the same periodicity as the DNA helix. Shown is a densitometer scan of the autoradiograph of a denaturing polyacrylamide electrophoresis gel on which was separated the products of cleavage by hydroxyl radical of the Herpes DNA molecule. Each peak in the pattern represents cleavage at a particular nucleotide of the Herpes DNA: the area of a peak gives the rate of attack at that site. The pattern goes through 14 sine-like oscillations in 147 bases, for a periodicity of $147/14 = 10.5$ base pairs per helical turn. Sequences that are bound by proteins that regulate transcription of the Herpes gene are indicated above the pattern. Adapted from (59) with permission (Copyright 1985 by the AAAS).

Mixed-sequence DNA acts as a straight rod over a dozen or more turns of the helix. In many biological milieus, though, DNA must bend, for example to be packaged within the head of a bacteriophage or to be wrapped around the histone octamers of chromatin. These bends occur through the agency of protein-DNA interactions. Remarkably, there are now examples of DNA molecules that are stable curved on their own. The structures of naturally bent DNA molecules will be important in understanding how DNA can be induced by proteins to bend.

The kinetoplast body of the trypanosomatids contains a particularly striking example of naturally curved DNA. Fragments of kinetoplast DNA (K-DNA) were first recognized to be bent because of their reduced mobility during electrophoresis on polyacrylamide gels (72). Electrophoresis is still the most widely used method for detecting bent DNA, but since it measures a global property of the DNA molecule, the details of DNA structure that cause bending are difficult to discern from such experiments.

Wu and Crothers used an elegant extension of the electrophoretic mobility experiment to map the center of the bend in *Leishmania tarentolae* kinetoplast DNA (73). They found a very suggestive DNA sequence at the bend locus. The sequence motif consisted of a short run of adenines, four or five in length, followed by five or six other nucleotides of apparently random sequence. This motif was repeated four times in the bent *Leishmania* DNA. Since poly(dA)•poly(dT) was known to adopt a conformation different from B-DNA, the reason for bending was suggested. The junctions between an adenine tract and the flanking sequence might cause discontinuities in the helix axis. In kinetoplast DNA these junctions would occur once every 10 base pairs (or one turn of the helix). The presumed discontinuities would thus add in phase in a plane and give an overall curvature to kinetoplast DNA.

Although it is known to be unusual, the structure of poly(dA)• poly(dT) is still not known with certainty. So besides having only a global picture of bent DNA from electrophoretic mobility studies, the details of the structure of an adenine tract that causes DNA to bend are unclear. We set out to use the chemistry of the hydroxyl radical to fill in some of the details of the conformations of the individual nucleotides in bent kinetoplast DNA.

By the time we started to think about this problem, Kitchin et al. had isolated a kinetoplast DNA fragment from *Crithidia fasciculata* that promised to be ideal for structural studies (74). The *Crithidia* K-DNA molecule contained 18 phased adenine tracts, and was shown to run very slowly during electrophoresis on polyacrylamide. Electron micrographs of this DNA molecule demonstrated conclusively its bent nature: a 200 base pair-long restriction fragment containing the 18 adenine tracts was shown to bend into a 360° open circle (75).

The hydroxyl radical cutting pattern of the *Crithidia* K-DNA fragment was the most unusual we had seen (60). Instead of a uniform degree of cleavage at each nucleotide, we found that the cutting frequency had a strikingly sinusoidal appearance (Figure 5, top). Each run of adenines was phased perfectly with the sine wave of the cutting pattern. The 3' adenine in each tract always was found at a minimum in the pattern. A smooth increase in cleavage frequency occurred from 3' to 5' along the adenine tracts, with the 5' adenines always near the maxima in the pattern.

The hydroxyl radical cleavage pattern immediately showed that the adenines in a tract were not equivalent. The deoxyribose of a 5' adenine was attacked at near normal frequency, while that of a 3' adenine was always cut to a much lower extent. The cutting pattern also told us that the intervening nucleotides were affected by the presence of an A tract. The cleavage frequency did not sharply return to normal after the 3' adenine, but "relaxed" over three or four nucleotides.

The key observation in our determination of the structural reason for DNA bending was the correlation in hydroxyl radical cutting frequency of the two strands (Figure 5, bottom). We found that the thymine-rich strand also had a sine wave-shaped cutting pattern, but one out of phase with the pattern on the adenine-rich strand. A particular adenine nucleotide was cut at a similar rate to the thymine that was one or two nucleotides in the 3' direction on the thymine-rich strand. This means that the structural feature that gives rise to a particular cutting frequency is shared not by an adenine and its base-paired thymine, but by the adenine and the thymine that is directly across the minor groove. This conclusion follows from the geometry of right-handed DNA, for which the shortest distance across the minor groove is from one nucleotide to the nucleotide on the opposite strand two or three nucleotides back.

We think that the most likely reason that the smooth decrease in cutting along an adenine tract is associated with the minor groove is that the minor groove narrows from the 5' adenine in a tract to the 3' adenine. Others have noted that poly(dA)•poly(dT) has a narrow minor groove compared to mixed sequence B-form DNA (76). A recent crystal structure of an oligonucleotide containing six consecutive adenines also shows a narrow minor groove (77). We conclude that this smooth narrowing and then opening up of the minor groove, repeated perfectly in phase with the DNA helical repeat, is related to the bending of kinetoplast DNA.

We have found that single adenine tracts also have a similar unusual hydroxyl radical cutting pattern (60). Variations in DNA conformation of this sort can therefore be detected by hydroxyl radical, at a resolution sufficient to give information on the structure of individual nucleotides.

Conclusions and Prospects

I have surveyed in this chapter a number of strategies for using metal complexes as tools in molecular biology. Metal complexes are especially

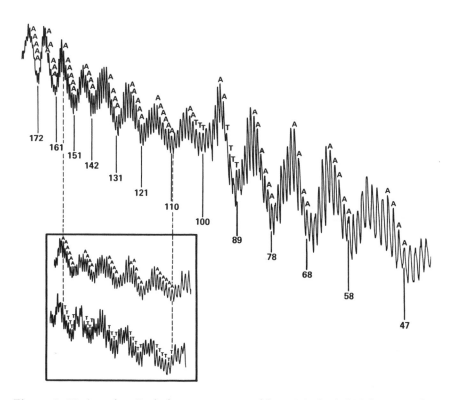

Figure 5. Hydroxyl radical cleavage pattern of bent DNA. A 210 base pair-long restriction fragment from the kinetoplast DNA of the trypanosomatid Crithidia fasciculata *was radioactively labeled and cleaved by the hydoxyl radical. Densitometer scans of gel autoradiographs are shown for the two strands. The adenine-rich strand (top) runs 5' to 3', left to right, while the thymine-rich complementary strand (bottom) runs 3' to 5', left to right. The tracts of adenines that have been associated with DNA bending are found to be phased precisely with the modulated cleavage pattern: the adenines at the 3' ends of tracts occur at the minima of the pattern, while the adenines at the 5' ends of adenine tracts are at the maxima. The thymine-rich strand also exhibits a modulated cleavage pattern, but one slightly out of phase from the pattern on the adenine-rich strand. These cleavage patterns have been interpreted to provide details of the unusual structure adopted by bent DNA (see text for details). Adapted from (60), with permission.*

well-suited to use as probes of unusual DNA structures, and as artificial enzymes for manipulating DNA. Metals have the advantage over their natural counterparts, the enzymes, that an arbitrarily high degree of sequence specificity can in principle be designed into the metal complex. There is little doubt that metal complexes will be of great importance as "molecular scalpels" for dissecting the genome into manageable pieces for mapping and sequencing.

The ability to design metal complexes to interact with specific, unusual, DNA structures is also in its infancy. A key question is to what extent these structures (hairpins, triple strands, Z-DNA, and the like) are involved in functioning biological systems. A definitive answer will come only from studies of DNA structure in living cells. Metal complexes will likely be widely used to approach this problem.

Acknowledgements

I wish to express my appreciation to Beth Dombroski and Amanda Burkhoff, the students who accomplished the studies in my laboratory that I described. I am grateful for financial support that was provided by a Research Career Development Award from the National Cancer Institute of the National Institutes of Health (CA 01208), a fellowship from the Alfred P. Sloan Foundation, and a Camille and Henry Dreyfus Teacher-Scholar award.

Literature Cited

1. Tullius, T. D. *Trends Biochem. Sci.* **1987,** *12,* 297–300.
2. *Platinum, Gold, and Other Chemotherapeutic Agents;* Lippard, S. J., Ed.; ACS Symposium Series No. 209; American Chemical Society: Washington, D. C., **1983.**
3. Jacob, F.; Monod, J. *J. Mol. Biol.* **1961,** *3,* 318–356.
4. Ptashne, M. "A Genetic Switch"; Cell Press & Blackwell Scientific Publications: Cambridge, MA & Palo Alto, CA, **1986.**
5. Watson, J. D.; Crick, F. H. C. *Nature* **1953,** *171,* 737–738.
6. Wells, R. D. *J. Biol. Chem.* **1988,** *263,* 1095–1098.
7. Wells, R. D.; Goodman, T. C.; Hillen, W.; Horn, G. T.; Klein, R. D.; Larson, J. E.; Muller, U. R.; Neuendorf, S. K.; Panayotatos, N.; Stirdivant, S. M. *Prog. Nucl. Acid Res. & Mol. Biol.* **1980,** *24,* 167–267.
8. McClarin, J. A.; Frederick, C. A.; Wang, B-C; Greene, P.; Boyer, H. W.; Grable, J.; Rosenberg, J. M. *Science* **1986,** *234,* 1526–1541.
9. Koudelka, G. B.; Harrison, S. C.; Ptashne, M. *Nature* **1987,** *326,* 886–891.
10. Aggarwal, A. K.; Rodgers, D. W.; Drottar, M.; Ptashne, M.; Harrison, S. C. *Science* **1988,** *242,* 899–907.

11. Dickerson, R. E.; Drew, H. R.; Conner, B. N.; Wing, R. M.; Fratini, A. V.; Kopka, M. L. *Science* **1982**, *216*, 475–485.

12. Pabo, C. O.; Sauer, R. T. *Ann. Rev. Biochem.* **1984**, *53*, 293–321.

13. Jordan, S. R.; Pabo, C. O. *Science* **1988**, *242*, 893–899.

14. Kaptein, R.; Boelens, R. M.; Scheek, R. M.; van Gunsteren, W. F. *Biochemistry* **1988**, *27*, 5389–5395.

15. von Hippel, P. H.; Bear, D. G.; Morgan, W. D.; McSwiggen, J. A. *Ann. Rev. Biochem.* **1984**, *53*, 389–446.

16. Serfling, E.; Jasin, M.; Shaffner, W. *Trends Genet.* **1985**, *1*, 224–230.

17. Tullius, T. D. *Ann. Rev. Biophys. Biophys. Chem.* **1989**, *18*, 213–237.

18. Maniatis, T.; Fritsch, E. F.; Sambrook, J. "Molecular Cloning–A Laboratory Manual"; Cold Spring Harbor Laboratory: Cold Spring Harbor, NY, **1982**.

19. Maxam, A. M.; Gilbert, W. *Meth. Enzymol.* **1980**, *65*, 499–560.

20. Lilley, D. M. J. *Proc. Natl. Acad. Sci. USA* **1980**, *77*, 6468–6472.

21. Barton, J. K. *Science* **1986**, *233*, 727–734.

22. Dervan, P. B. *Science* **1986**, *232*, 464–471.

23. Youngquist, R. S.; Dervan, P. B. *J. Am. Chem. Soc.* **1987**, *109*, 7654–7566.

24. Moser, H. E.; Dervan, P. B. *Science* **1987**, *238*, 645–650.

25. Strobel, S. A.; Moser, H. E.; Dervan, P. B. *J. Am. Chem. Soc.* **1988**, *110*, 7927–7929.

26. Sigman, D. A.; Graham, D. R.; D'Aurora, V.; Stern, A. M. *J. Biol. Chem.* **1979**, *254*, 12269–12271.

27. Sigman, D. S. *Acc. Chem. Res.* **1986**, *19*, 180–186.

28. Spassky, A.; Sigman, D. S. *Biochemistry* **1985**, *24*, 8050–8056.

29. Kuwabara, M. D.; Sigman, D. S. *Biochemistry* **1987**, *26*, 7234–7238.

30. Hertzberg, R. P.; Dervan, P. B. *J. Am. Chem. Soc.* **1982**, *104*, 313–314.

31. Hertzberg, R. P.; Dervan, P. B. *Biochemistry* **1984**, *23*, 3934–3945.

32. Van Dyke, M. W.; Dervan, P. B. *Nucleic Acids Res.* **1983**, *11*, 5555–5567.

33. Van Dyke, M. W.; Dervan, P. B. *Cold Spring Harbor Symp. Quant. Biol.* **1983**, *XLVII*, 347–353.

34. Van Dyke, M. M.; Dervan, P. B. *Science* **1984**, *225*, 1122–1127.

35. Sawodogo, M.; Roeder, R. G. *Cell* **1985**, *43*, 165–175.

36. Tullius, T. D.; Dombroski, B. A.; Churchill, M. E. A.; Kam, L. *Meth. Enzymol.* **1987**, *155*, 537–558.

37. Schultz, P. G.; Dervan, P. B. *J. Am. Chem. Soc.* **1983**, *105*, 7748–7750.

38. Youngquist, R. S.; Dervan, P. B. *Proc. Natl. Acad. Sci. USA* **1985**, *82*, 2565–2569.

39. Schultz, P. G.; Taylor, J. S.; Dervan, P. B. *J. Am. Chem. Soc.* **1982**, *104*, 6861.

40. Dreyer, G. B.; Dervan, P. B. *Proc. Natl. Acad. Sci. USA* **1985**, *82*, 968–972.
41. Chen, C.-H B.; Sigman, D. S. *Proc. Natl. Acad. Sci. USA* **1986**, *83*, 7147–7151.
42. Francois, J. C.; Saison-Behmoaras, T.; Chassignol, M.; Thuong, N. T.; Sun, J. S.; Helene, C. *Biochemistry* **1988**, *27*, 2272–2276.
43. Le Doan, T.; Perrouault, L.; Helene, C.; Chassignol, M.; Thuong, N. T. *Biochemistry* **1986**, *25*, 6736–6739.
44. Trung, L. D.; Perrouault, L.; Chassignol, M.; Nguyen, T. T.; Helene, C. *Nucleic Acids Res.* **1987**, *15*, 8643–8659.
45. Boutorin, A.; Vlassov, V. V.; Koyakov, S. A.; Kutiavin, I. V.; Podiminoyin, M. A. *FEBS Lett.* **1984**, *172*, 46–48.
46. Chu, B. C.; Orgel, L. E. *Proc. Natl. Acad. Sci. USA* **1985**, *82*, 963–967.
47. Boidot-Forget, M.; Thuong, N. T.; Chassignol, M.; Helene, C. *C. R. Acad. Sci., Ser. 2* **1986**, *302*, 75–80.
48. Sluka, J. P.; Horvath, S. J.; Bruist, M. F.; Simon, M. I.; Dervan, P. B. *Science* **1987**, *238*, 1129–1132.
49. Mack, D. P.; Iverson, B. L.; Dervan, P. B. *J. Am. Chem. Soc.* **1988**, *110*, 7572–7574.
50. Chen, C-H.; Sigman, D. S. *Science* **1987**, *237*, 1197–1201.
51. Barton, J. K. *J. Biomol. Struct. Dyn.* **1983**, *1*, 621–632.
52. D'Andrea, A. D.; Haseltine, W. A. *Proc. Natl. Acad. Sci. USA* **1978**, *75*, 3608–3612.
53. Barton, J. K.; Dannenberg, J. J.; Raphael, A. L. *J. Am. Chem. Soc.* **1982**, *104*, 4967–4969.
54. Barton, J. K.; Basile, L. A.; Danishefsky, A.; Alexandrescu, A. *Proc. Natl. Acad. Sci. USA* **1984**, *81*, 1961–1965.
55. Barton, J. K.; Danishefsky, A.; Goldberg, J. *J. Am. Chem. Soc.* **1984**, *106*, 2172–2176.
56. Barton, J. K.; Raphael, A. L. *J. Am. Chem. Soc.* **1984**, *106*, 2466–2468.
57. Mei, H. Y.; Barton, J. K. *Proc. Natl. Acad. Sci. USA* **1988**, *85*, 1339–1343.
58. Lippard, S. J. *Acc. Chem. Res.* **1978**, *11*, 211–217.
59. Tullius, T. D.; Dombroski, B. A. *Science* **1985**, *230*, 679–681.
60. Burkhoff, A. M.; Tullius, T. D. *Cell* **1987**, *48*, 935–943.
61. Burkhoff, A. M.; Tullius, T. D. *Nature* **1988**, *331*, 455–457.
62. Walling, C. *Acc. Chem. Res.* **1975**, *8*, 125–131.
63. Fenton, H. J. H. *J. Chem. Soc.* **1894**, *65*, 899.
64. Udenfriend, S.; Clark, C. T.; Axelrod, J.; Brodie, B. B. *Federation Proc.* **1953**, *11*, 731–739.
65. Rhodes, D.; Klug, A. *Nature* **1980**, *286*, 573–578.
66. Liu, L. F.; Wang, J. C. *Cell* **1978**, *15*, 979–984.
67. Lutter, L. C. *Nucleic Acids Res.* **1979**, *6*, 41–56.

68. Drew, H. R.; Travers, A. A. *Cell* **1984**, *37*, 491–502.
69. McKnight, S. L.; Kingsbury, R. *Science* **1982**, *217*, 316–324.
70. McKnight, S.; Tjian, R. *Cell* **1986**, *46*, 795–805.
71. Ptashne, M. *Nature* **1986**, *322*, 697–701.
72. Marini, J. C.; Levene, S. D.; Crothers, D. M.; Englund, P. T. *Proc. Natl. Acad. Sci.USA* **1982**, *79*, 7664–7668.
73. Wu, H-M; Crothers, D. M. *Nature* **1984**, *308*, 509–513.
74. Kitchin, P. A.; Klein, V. A.; Ryan, K. A.; Gann, K. L.; Rauch, C. A.; Kang, D. S.; Wells, R. D.; Englund, P. T. *J. Biol. Chem.* **1986**, *261*, 11302–11309.
75. Griffith, J.; Bleyman, M.; Rauch, C. A.; Kitchin, P. A.; Englund, P. T. *Cell* **1986**, *46*, 717–724.
76. Alexeev, D.; Lipanov, A.; Skuratovskii, I. *Nature* **1987**, *325*, 821–823.
77. Nelson, H. C. M.; Finch, J. T.; Bonaventura, F. L.; Klug, A. *Nature* **1987**, *330*, 221–226.

RECEIVED May 30, 1989

Chapter 2

Chemical Nuclease Activity of 1,10-Phenanthrolinecopper

A New Reagent for Molecular Biology

David S. Sigman and Chi-hong B. Chen

Department of Biological Chemistry, School of Medicine and Molecular Biology Institute, University of California, Los Angeles, CA 90024

1,10-Phenanthroline-copper ion is an efficient chemical nuclease which makes single-stranded nicks in DNA at physiological pH and temperature. In addition to serving as a reliable footprinting reagent, it is useful for defining functionally-important sequence-dependent conformational variability of DNA and protein-induced structural changes in DNA. The intrinsic specificity of the reagent can be overridden by linking it to deoxyoligonucleotides or DNA-binding proteins. Site-specific scission, reflecting the binding site of the targeting ligand, is then observed.

Naturally-occurring nucleases cleave the phosphodiester backbone of DNA by hydrolysis (1). The products are either 3' phosphates and free 5' ends or 5' phosphates and free 3' ends. Recently, chemical nucleases have been described which cleave DNA under physiological conditions by oxidatively degrading the deoxyribose moiety. These nucleolytic activities include 1,10-phenanthroline-copper (2), ferrous EDTA as a free coordination complex or linked to an ethidium bromide derivative (3, 4), iron porphyrins (5, 6), and the ruthenium and cobalt complexes of 4,7-diphenyl-1,10-phenanthroline (7). Chemical methods for nicking DNA have several uses. They include probing the sequence-dependent conformational variability of DNA, identifying the binding sites of DNA ligands, and serving as the catalytic entity in the design of artificial sequence-specific nucleases.

Although the primary site of attack of these reagents is the deoxy-ribose, the chemical nature of the oxidative species is not the same in all cases. For example, ferrous EDTA appears to function by chemically generating diffusible hydroxyl radicals in the presence of oxygen and a reducing agent (8). Once produced, these hydroxyl radicals must degrade DNA by a mechanism which parallels that of hydroxyl radicals generated by cobalt-60 radiation (9). The light-activated scission activity of the octa-hedral complexes of 4,7-diphenyl-1,10-phenanthroline may involve sing-

0097–6156/89/0402–0024$06.75/0

let oxygen (7). Although the detailed chemistry of the scission reaction of the iron porphyrin complexes has not been fully investigated, the 1,10-phenanthroline-copper reaction may involve a copper-oxo species, possibly reminiscent of the species involved in the mechanism of action of the cytochrome P-450 systems (10). With the possible exception of octahedral complexes of 4,7-diphenyl-1,10-phenanthroline, all these chemical nuclease activities involve attack on the deoxyribose moiety. As a result, their cutting reaction is not expected to be base specific; oxidation of the deoxyribose moiety is not dependent on the base attached to the deoxyribose moiety attacked. This feature sharply distinguishes these chemical reagents from those used in Maxam-Gilbert sequencing reactions (11). These, of course, are base specific, and lead to the formation of apurinic or apyrimidinic sites at which strand scission occurs following treatment with piperidine.

Mechanism of DNA Scission by 1,10-Phenanthroline-Copper Ion

This chapter will focus on the nuclease activity of the 1,10-phenanthroline-copper complex (OP-Cu) which has been extensively studied in our laboratory following its discovery in 1979 (2). Initially identified as the process responsible for the observed 1,10-phenanthroline inhibition of a variety of RNA and DNA polymerases (12, 13), it was the first chemical nuclease activity described (2). In this reaction, there are two essential coreactants, the 2:1 1,10-phenanthroline-cuprous complex and hydrogen peroxide (2, 14). The kinetic scheme of the reaction is summarized in Figure 1 and is valid whether superoxide or thiol is used as reductant. The first step is the reduction of 1,10-phenanthroline-cupric ion $[(OP)_2Cu^{2+}]$ to form the 2:1 1,10-phenanthroline-cuprous complex $[(OP)_2Cu^+]$ in solution. This tetrahedral complex binds to DNA to form a central intermediate through which the reaction is funnelled. Hydrogen peroxide, which can be added exogenously or generated *in situ* by oxidation of the cuprous complex under aerobic conditions, then reacts with the DNA-bound 1,10-phenanthroline cuprous complex (14, 15). Formally analogous to the Fenton reaction, this one-electron oxidation of the cuprous ion leads to a reactive species which can be written either as a hydroxyl radical coordinated to a cupric ion, or as a copper-oxene structure analogous to the ferryl [iron(IV)]-radical cation porphyrin structure of likely importance in the the reaction mechanism of cytochrome P-450 (10). Possibly, the unique reactivity of the 1,10-phenanthroline nucleus in the DNA cleavage reaction can be attributed to the transfer of charge from the phenanthroline to the metal ion. The cognate ligands 2,2-bipyridine and 2,2',2''-terpyridine, whose copper complexes are inactive in DNA scission, may be unable to form stable copper-oxo species in a parallel manner. The copper chelates of these ligands, lacking a central ring, may also not form stable complexes with DNA.

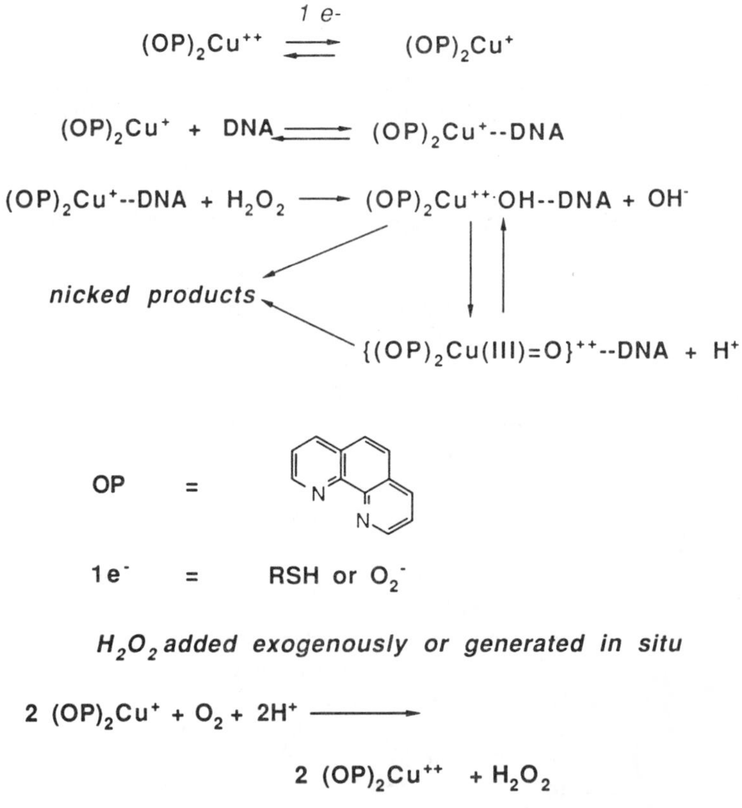

$$(OP)_2Cu^{++} \xrightleftharpoons{1\ e^-} (OP)_2Cu^+$$

$$(OP)_2Cu^+ + DNA \rightleftharpoons (OP)_2Cu^+\text{--}DNA$$

$$(OP)_2Cu^+\text{--}DNA + H_2O_2 \longrightarrow (OP)_2Cu^{++}\cdot OH\text{--}DNA + OH^-$$

nicked products

$$\{(OP)_2Cu(III)=O\}^{++}\text{--}DNA + H^+$$

OP   =

1e⁻   =   RSH or O_2^-

H_2O_2 added exogenously or generated in situ

$$2\ (OP)_2Cu^+ + O_2 + 2H^+ \longrightarrow$$

$$2\ (OP)_2Cu^{++} + H_2O_2$$

Figure 1. Proposed kinetic scheme for the nuclease activity of 1,10-phenan-throline-copper ion. The cuprous complex is formed in situ by the reduction of the cupric complex either by thiol or superoxide. Hydrogen peroxide can be formed from the oxidation of the diffusible cuprous complex by molecular oxygen. Alternatively, it may be added exogenously.

Detailed product analysis of the scission reaction has demonstrated that the reaction scheme summarized in Figure 2 is valid and represents, on the average, as much as 90% of the phosphodiester bond cleavage activity (15–17). Greater than 0.9 mole of 5-methylene-2-furanone has been isolated per mole of strand break in the cleavage of poly(dAT) (17). Further confirmation of this scheme is provided by the isolation of unmodified base, and the demonstration that metastable species are detectable if 5'-labelled DNA is used as substrate (16). As would be anticipated from the mechanism as written, no transient species at the 5' end is detected when 3'-labelled DNA is used as substrate. There are two important mechanistic conclusions which arise from an understanding of the detailed chemistry for the reaction. First, it is unlikely that diffusible hydroxyl radical is responsible for the strand scission. In this case, equivalent yields of 3' phosphoglycolates and 3' phosphates would be expected. 3' phosphoglycolates would be generated from attack at C-4 of the deoxyribose, and have been observed in the scission of DNA by methidiumpropyl-EDTA•iron(II) and hydroxyl radicals generated from Co-60 radiation (8, 9). Secondly, the reaction mechanism demands that attack by OP-Cu proceeds from a binding site of the coordination complex within the minor groove. The C-1 hydrogen is deeply recessed within the minor groove (18) and could only be accessible if the copper-oxo species binds close to it.

Using DNA-Ligand Complexes of Known Structure to Verify Minor Groove Attack

Further evidence for the binding of the coordination complex to the minor groove can be derived from experiments using OP-Cu as a footprinting reagent to examine the binding of two DNA-ligand complexes of known structure, the complex of netropsin with the self-complementing dodecamer 5'-d(CGCGAATTCGCG)-3' (19), and the complex of *Eco*RI with this same deoxyoligonucleotide. The crystallographic results demonstrate that netropsin binds to the central AATT residues in the minor groove (20). Consistent with the hypothesis that the coordination complex attacks from this domain, DNA scission is blocked in this region. In contrast, the restriction enzyme *Eco*RI binds exclusively to the major groove, leaving the minor groove accessible to solvent (20); in this case, only minor perturbations in the OP-Cu cutting pattern are apparent (16).

Secondary Structure Specificity

Initial evidence for the formation of a kinetically important intermediate complex of the coordination complex with DNA was found in the specificity exhibited by OP-Cu in cutting DNA's of different secondary structure (21). For example, B-DNA proved to be the most susceptible helical

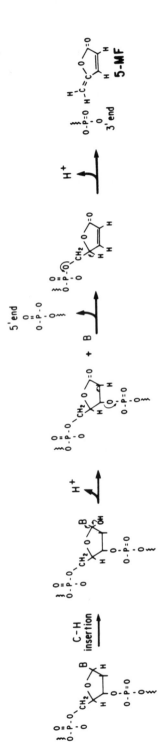

Figure 2. Proposed reaction mechanism. Initial site of attack by the oxidative species is C-1 of the deoxyribose. The reaction mechanism is supported by the isolation of free bases, 5' and 3' phosphorylated termini, and 5-methylene-furanone. A transient intermediate has been identified on the 3' terminus but not the 5' terminus.

structure to cleavage. A-DNA was cut about 1/4 to 1/3 slower, and Z-DNA and non-self-complementary single-stranded DNA's were not measurably cleaved (*21*). The efficiency of the reaction clearly depends on the availability of a stable binding site for the coordination complex. This accounts not only for the specificity of the reagent for different secondary structures but for the observed reactivity with naturally occurring DNA's.

Sequence-Dependent Reactivity in B-DNA

OP-Cu shows pronounced sequence dependence in its reaction with DNA. The importance of primary sequence in the cleavage chemistry is underscored by comparing the cleavage pattern of a portion of the *lac* operator in a 186 base pair restriction fragment to that of a deoxyoligo-nucleotide comprising sequence positions +1 to +21 of the *lac* transcription unit (Figure 3). This Figure summarizes several important features of the reactivity of OP-Cu with B-DNA. Examination of the digestion reveals that any of the four bases can be present in a hyperreactive or hyporeactive sequence. Analysis of the digestion patterns of a variety of DNA's has revealed that C is slightly disfavored relative to A and T, and that G is slightly favored as a cutting site with respect to A and T (*22*). This is unlikely to be reflective of a difference in the intrinsic reactivity of the deoxyribose, but instead may be due to the stability and reactivity of the complex between OP-Cu and the DNA near these sites. The second feature that is apparent from this Figure is that strong cutting sites present on one strand are associated with a series of strong cutting sites offset in the 3' direction on the opposite strand. This correlation of cutting is consistent with the minor groove binding of the reagents (*22, 23*). The similarity of the cutting of the deoxyoligonucleotide and the same sequence in a longer DNA strand strongly suggests that the local primary sequence is the major determinant governing cutting efficiency. The nucleotide 5' to the cutting site appears to be the primary determinant in governing the cutting, although the nucleotide 3' to the cutting site has some influence as well.

The sequence dependence of the reactivity of OP-Cu is one major difference between it and the chemical nuclease activities of the iron EDTA reagents. Neither methidiumpropyl-EDTA•Fe(II) [MPE], which reacts with DNA (*8*) after first intercalating, nor ferrous EDTA (*4*), which does not have high affinity for DNA, shows comparable sequence specificity in their digestion pattern. For example, the digestion of a fragment of DNA containing the region of transcription termination of the μ_m immunoglobulin gene by OP-Cu, DNase I, and MPE is compared in Figure 4 (*24*). MPE shows much less variation in its cleavage, although it is readily apparent that a completely even ladder is not generated by this reagent. The OP-Cu pattern is intermediate in its variability of cutting between MPE and DNase I.

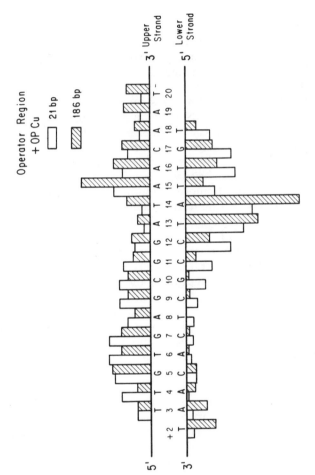

Figure 3. Comparison of the relative rate constants for the OP-Cu digestion of the operator region in 21 bp and 186 bp DNA fragments. Shaded columns represent digestion of the 186 bp EcoRI restriction fragment. Unshaded columns represent digestion of 21 bp long chemically synthesized oligonucleotides. The upper strand is the non-template strand.

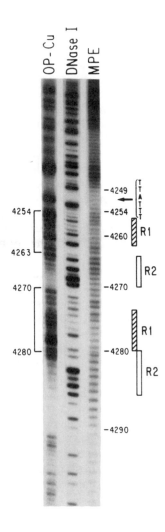

Figure 4. Comparison of the scission pattern of OP-Cu, MPE and DNase I. Nuclease cleavage patterns of the coding strand for the μ_m poly(A) addition region of the mu + delta gene. R1 and R2 are direct sequence repeats with characteristically different reactivity with OP-Cu and DNase I.

Biochemical Phenotype Associated with the Sequence-Dependent Conformational Variability of DNA

Are there any insights associated with this sequence-dependent cutting? There are two contexts where this aspect of the reaction has proven to be useful. In one, the digestion patterns of the wild type *lac* promoter was compared to that of the cyclic AMP-independent UV-5 promoter. The difference between these two promoters is at positions –9 and –8, residues within the promoter-conserved sequence of the Pribnow box (*25, 26*). As indicated in Figure 5, these two base changes dramatically alter the OP-Cu scission pattern. The mutational changes associated with the increased affinity of RNA polymerase to the promoter must therefore also be accompanied by an alteration of the minor groove geometry where OP-Cu binds. Enhanced cutting could be due to one or both of the following aspects. Since the binding step precedes the scission reaction, enhanced reactivity could be due to a thermodynamically more stable complex with the UV-5 promoter than with the wild-type promoter–a K_m effect. Alternatively, the stability of the intermediate complex of the coordination complex with DNA may be unchanged, but the mutational change might improve the orientation of the copper-oxo species to the C-1 of the deoxyribose–a V_{max} effect.

The digestion pattern of OP-Cu of the enhancer sequence of the κ immunoglobulin light chain gene is presented in bar graph form in Figure 6A (*27*). The identical sequence is present in the transcription control region in the long terminal repeat (LTR) of the human immuno-deficiency virus. Our studies have indicated that these sequences have different affinities for transcription factors produced in phorbol ester-stimulated HeLa cells. How can the same sequence differ in binding the same protein factor? One possible explanation is that the sequences which flank the recognition sequence of these transcription factors influence the DNA conformation. This is anticipated from the analysis of the primary sequence specificity of the OP-Cu reagent in which 5' flanking sequences have a major influence on digestion rates. In Figure 6B the cutting pattern of the same DNA sequence in the context of the HIV control region is presented, revealing substantial differences from that of the identical sequence in the κ light chain control region (*27*).

Gel Retardation / OP-Cu Footprinting

Chemical nucleases are small and diffusible in solid supports. We have utilized this property to develop a methodology which enables us to couple OP-Cu footprinting with the widely used gel retardation assay to study DNA-protein interactions (*28*). In gel retardation assays, the binding of a protein to a labeled restriction fragment alters the migration of the DNA in a non-denaturing acrylamide gel (*29, 30*). Even in a complex

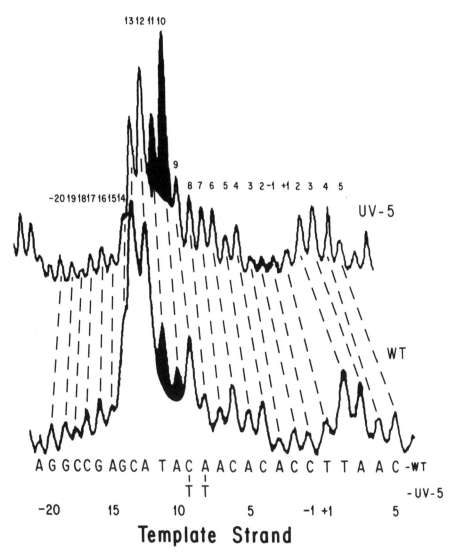

Figure 5. Comparison of OP-Cu scission patterns of UV-5 and wt lac *operators. The sequence of the wt DNA is indicated. The nucleotide changes of the UV-5 mutation are shown below.*

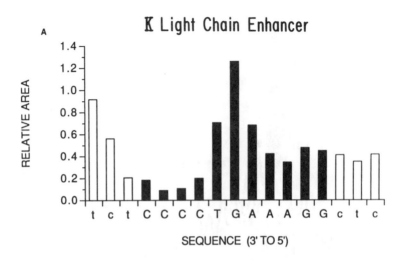

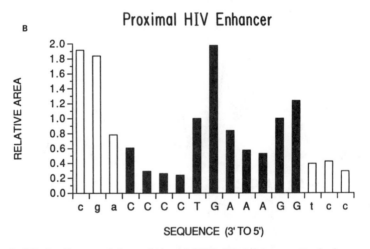

Figure 6. OP-Cu digests of the κ (A) and HIV (B) enhancers. Both the proximal 4G HIV and κ enhancer fragments were digested with OP-Cu in the absence of added protein and subjected to autoradiography and densitometry. The 5' base influences the cutting pattern within the recognition sequence.

mixture of proteins, it is possible to separate a protein which binds to a labeled DNA. To identify the sequence-specific contacts between the protein and DNA, the footprinting reaction with OP-Cu can be carried out directly in the gel (28). The deoxyoligonucleotide products can then be directly eluted from the acrylamide of the retardation gel and analyzed on sequencing gels to obtain the sequence domain of the binding site of the protein-DNA complex detected in the retardation assay. The sequencing gels presented in Figure 7 indicate that the footprint for the binding of the *lac* repressor to the *lac* operator obtained by this method is the same as that obtained by carrying out footprints in solution using either OP-Cu or DNase I as the nucleolytic agents (28, 31).

Footprints of unstable complexes that are difficult to detect in solution can be obtained by combining these two methodologies. There are two reason for this. First, unbound DNA is separated from the DNA-protein complex and therefore the background cutting of unbound DNA is greatly reduced. Secondly, the gel matrix inhibits the dissociation rate of the protein from its binding site on DNA. This reduced dissociation rate leads to a more efficient blockage of the access of the OP-Cu to the DNA. As a result, one weak, but biochemically significant, interaction that we were able to detect was the binding of a protein factor near the polyadenylation signal of the immunoglobulin μ_m transcription locus (24). The factor-binding site on the DNA is located 5' to the polyadenylation signal AATAAA, and includes the 15 nucleotide-long A/T-rich palindrome 5'-CGTAAACAAATGTC-3' (Figure 8). This type of palindromic binding site exhibits orientation-dependent activity consistent with the reported properties of RNA polymerase II termination signals. This binding site is followed by two sets of directly repeated DNA sequences with different helical conformations as revealed by their reactivity with OP-Cu. The close proximity of these features to the signals for μ_m mRNA processing may reflect a linkage of the processes of developmentally regulated μ_m polyadenylation and transcription termination (24).

This methodology has also been used by us to study the binding of RNA polymerase to the *lac* UV-5 promoter (28). One of the unexpected benefits of OP-Cu footprinting in solution has been its ability to detect transcriptionally competent initiation complexes of *E. coli* RNA polymerase with different bacterial promoters by the appearance of hyperreactive cleavage sites on the template strand (26). These bands correspond to sequence positions that are single-stranded in the enzyme-substrate complex composed of RNA polymerase and the UV-5 promoter. However, following the separation of polymerase-promoter complexes using a gel retardation assay, the hyperreactive bands were not observed unless magnesium ion was added (Figure 9). Their appearance upon the addition of magnesium ion indicates that this ion is essential for the enzyme-induced melting of the double stranded DNA, but not for the stability of the polymerase-DNA complex. The binding of a protein to a DNA molecule can be distinguished from protein-induced changes in DNA by

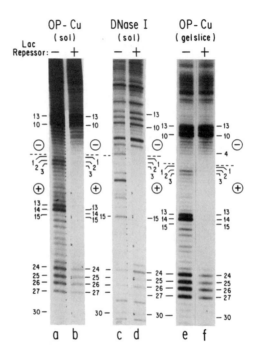

Figure 7. Comparison of the OP-Cu and DNase I footprints of the lac *repressor-operator complex in solution, with digestion of the complex in an acrylamide matrix following separation in a gel retardation assay. The footprinting reaction was carried out in an acrylamide plug that was excised from the gel. The 1:1 protein-DNA complex was isolated from an incubation mixture containing 9.4 nM L8-UV-5 DNA and 19 nM* lac *repressor. Lanes a and b, OP-Cu digestion in solution: a, free DNA; b, repressor-DNA complex; Lanes c and d, DNase I digestion in solution; c, free DNA: d, repressor-DNA complex; lanes e and f, OP-Cu digestion within the gel slice: e, unbound DNA from gel retardation assay; f, 1:1 repressor-DNA complex from gel retardation assay.*

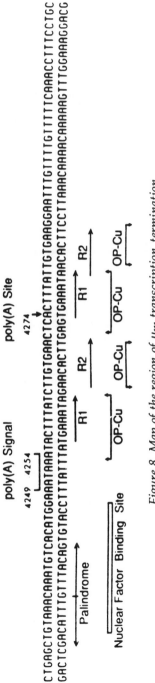

Figure 8. Map of the region of μm transcription termination.

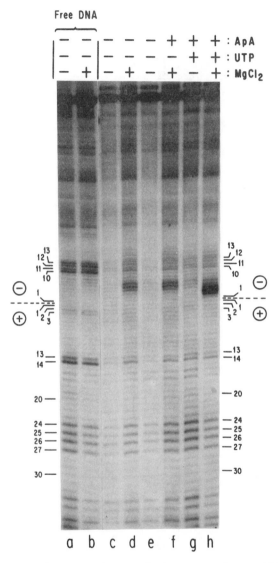

Figure 9. Digestion of L8 UV-5/RNA polymerase complexes within a gel slice. A gel slice containing unbound DNA was excised from a retardation gel in which the RNA polymerase-lac fragment was incubated with magnesium ion, or with magnesium ion, ApA, and UTP. The gel slices containing the DNA and DNA-protein complexes were immersed in 50 mM Tris•HCl buffer, pH 8.0, containing the identical added components. When included, [Mg²⁺]= 10 mM, [ApA]= 500 μM, and [UTP]= 50 μM. Lanes a–d were obtained from magnesium-only incubation; lanes e–h were obtained from magnesium ion, ApA, and UTP incubation. Digestion patterns of free DNA are the same for all incubation conditions.

coupling these two methodologies into one procedure. When an incomplete complement of ribonucleoside triphosphates are added to the footprinting reaction and short RNA's synthesized, these hyperreactive bands are displaced along the template strand (32). This observation is significant because it demonstrates that DNA-protein complexes remain functional within the acrylamide matrix, and induced changes in DNA structure associated with function can be followed.

Targeting OP-Cu Cutting

The unsubstituted 2:1 1,10-phenanthroline-cuprous complex shows valuable specificity in its scission chemistry which is a result of its formation of a DNA-coordination complex during the course of the reaction. We investigated whether it was possible to override the intrinsic specificity of the free coordination chemistry and target the scission chemistry by linking it to a carrier ligand.

Oligonucleotide-Directed Scission. The first example of the targeted scission of OP-Cu utilized deoxyoligonucleotides as carriers (33). The 1,10-phenanthroline moiety was linked to the oligonucleotide using chemistry developed by Chu, Wahl, and Orgel (34). It involves the formation of an intermediate phosphoimidazolide using a water soluble carbodiimide as a coupling reagent. This linkage reacts with free amino groups (Figure 10) and can therefore be derivatized with 5-glycyl-amido-1,10-phenanthroline. Using this chemistry, any phosphorylated terminus of a deoxy- or ribo-oligonucleotide can be transformed into a carrier for the 1,10-phenanthroline moiety. An alternate method involves phosphorylating the RNA or DNA with ATP-γ-S to form a thiophosphate terminus which can then be alkylated with 5-iodoacetyl-1,10-phenanthroline.

The initial deoxyoligonucleotide derivatized in our laboratory was a 21-mer with a sequence complementary to positions 1–21 of *lac* mRNA. Site directed scission of a complementary DNA could be readily detected after hybridization. Two DNA's were used as targets. The first was single-stranded M13mp8 whose site-specific cutting was assayed by primer extension (35). A labeled restriction fragment containing the control region of the *lac* operon was also used as a target. Both assays demonstrated the same cutting pattern. The most prominent cutting sites were at sequence position 19–24, as indicated in Figure 11. If an RNA of identical sequence is used as a target for this site-specific scission reaction, parallel kinetics and specificity of scission are obtained (Figure 11). RNA and DNA strands, when constrained to the same conformation, appear to be equally sensitive to oxidative degradation. In view of the different intrinsic reactivities of deoxy- and ribo-oligonucleotides with respect to acid-catalyzed depurination (36), it was uncertain if they would be equivalently reactive

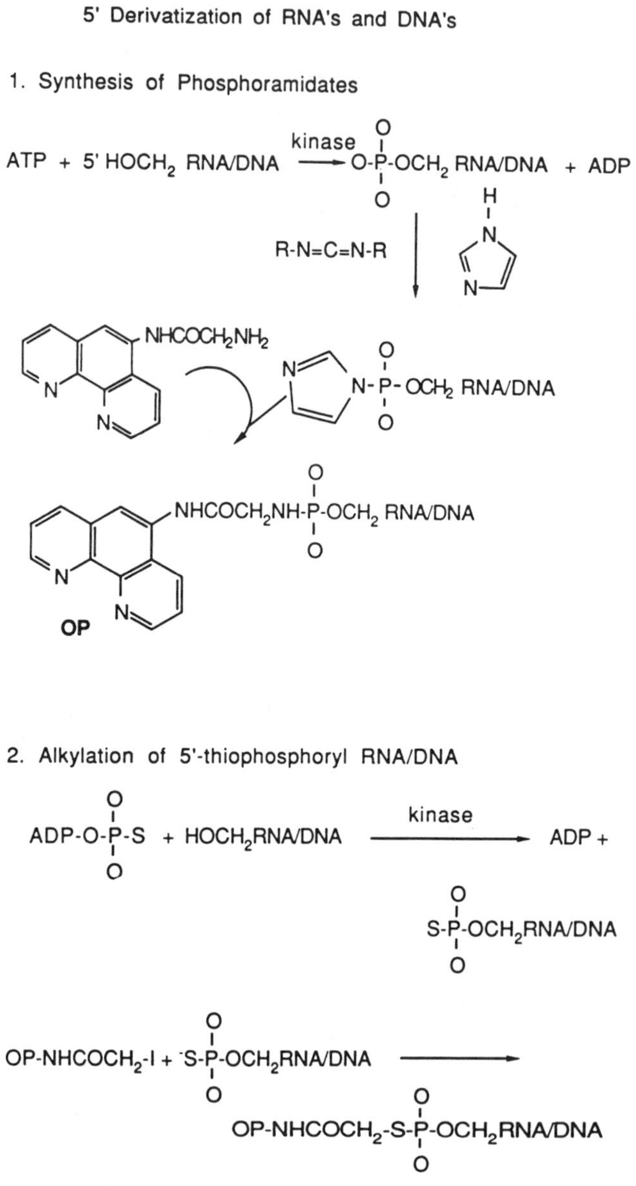
5' Derivatization of RNA's and DNA's
1. Synthesis of Phosphoramidates
ATP + 5' HOCH2 RNA/DNA
kinase
O-P-OCH2 RNA/DNA + ADP
R-N=C=N-R
NHCOCH2NH2
N-P-OCH2 RNA/DNA
NHCOCH2NH-P-OCH2 RNA/DNA
OP
2. Alkylation of 5'-thiophosphoryl RNA/DNA
ADP-O-P-S + HOCH2RNA/DNA
kinase
ADP +
S-P-OCH2RNA/DNA
OP-NHCOCH2-I + -S-P-OCH2RNA/DNA
OP-NHCOCH2-S-P-OCH2RNA/DNA

Figure 10. Methods of coupling 1,10-phenanthroline to RNA and DNA.

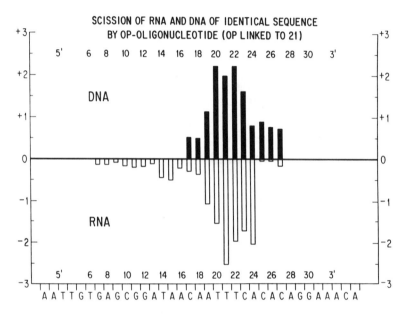

Figure 11. *Comparison of the scission of RNA and DNA with the sequence of the upper strand of the* E. coli lac *control region extending from positions –18 to + 63 using the reagent with the following structure:*

to a reagent whose primary site of attack would be the C-1 of the ribose moiety.

Initial studies of the nuclease activity of 1,10-phenanthroline copper complex revealed that the 2:1 1,10-phenanthroline copper complex had substantially greater efficiency than the 1:1 1,10-phenanthroline-copper complex (14). This conclusion was based on the demonstration that in the presence of 7 mM H_2O_2 and 7 mM mercaptopropionic acid, a solution containing 10 µM OP and 1 µM Cu^{2+} was efficient in the depolymerization of poly (dA-T), whereas a solution with 10 µM Cu^{2+} and 1 µM OP was very inefficient. The reason for the greater reactivity of the 2:1 complex is unclear. Possibly, both ligands are essential for the binding of the coordination complex to the DNA; alternatively, they are essential for the reactivity of the copper-oxo species. The demonstration that the OP-Cu reaction can be targeted by deoxyoligonucleotides and proteins suggests that a 1:1 OP-Cu complex has nucleolytic activity.

Protein Targeting of DNA Scission. Targeting the nuclease activity of 1,10-phenanthroline-copper can also be achieved using DNA binding proteins as the carrier ligand (37). Since DNA binding proteins frequently use recognition sequences as long as 20 bp long, semisynthetic nucleases generated from them would be infrequent cutters and possibly find use in chromosomal mapping studies. In addition, the advantage of using a protein as the recognition element in guiding the cleavage reaction is that it reads double stranded DNA with high affinity. The highly specific cutting inherent in using a single strand of RNA or DNA as a carrier for the nuclease activity depends on the development of hybridization procedures which would be applicable to complex, high molecular weight DNA.

The *E. coli* trp repressor was the initial protein chosen to determine if a DNA binding protein can be converted into a site-specific nuclease, because it is small, stable, of known structure, and readily available through the generosity of Professor Robert Gunsalus of the Department of Microbiology at UCLA (38, 39). Any protein can be modified using a simple protocol in which lysyl residues are first modified with iminothiolane (40) to generate sulfhydryl groups, which can then be derivatized with 5-iodoacetyl-1,10-phenanthroline (Figure 12).

Four 1,10-phenanthroline moieties were added per subunit of the trp repressor. Although the protein contains four lysyl residues, it is also possible that the N-terminal amino group is a site of modification as well. The potential problem of the approach outlined for the generation of protein-directed scission reagents is that the chemical modification of the lysyl residues may destroy site-specific binding. The DNase footprinting results summarized in Figure 13 indicate that this is not a problem in the present case. The modified trp repressor binds to the operator site of the *aroH* gene with roughly the same affinity as the unmodified repressor. But most significantly, upon the addition of cupric ion and thiol to the protein-DNA complex, DNA scission is observed within the recognition

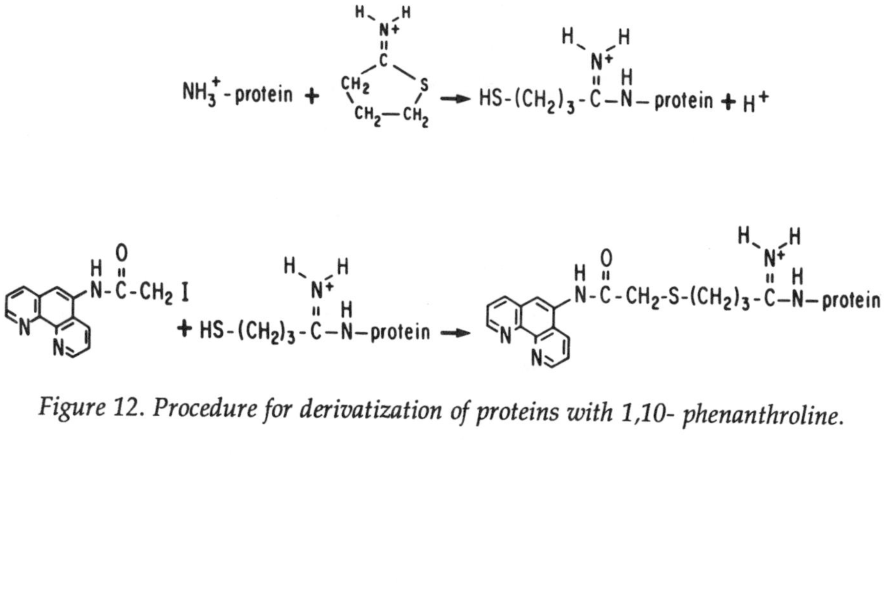

Figure 12. Procedure for derivatization of proteins with 1,10- phenanthroline.

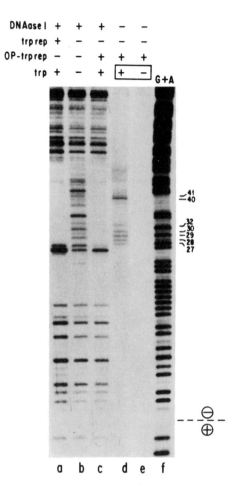

Figure 13. Binding and scission of the aroH *operator with OP-trp repressor. Template strand was 5'-labelled. Lanes a to c: DNase I footprints of the binding of trp repressor and OP-trp repressor to the 5' labelled template strand of* aroH. *Lane a, native trp repressor; lane b, control DNase I digestion; lane c, OP-trp repressor; lane d, OP-trp repressor with L-tryptophan, Cu²⁺, and 3-mercaptopropionic acid; lane e, same as lane d but lacking L-tryptophan; lane f, G + A sequencing lane.*

sequence. Important proof that the cutting reaction involves the formation of a protein-DNA complex is that it is absolutely dependent on the presence of L-tryptophan. Tryptophan acts as corepressor of the enzymes essential for its biosynthesis. The *E. coli* trp repressor does not bind to DNA unless tryptophan is bound to it and able to organize the protein into a conformation capable of high affinity interaction with the DNA (*39, 41*).

The modified trp repressor is able to nick the operator site which regulates the transcription of the *trp EDCBA* genes. This operator site shows sequence homology with that of the *aroH* operator (*42*). Interesting differences are apparent in the cutting patterns of the two operators. When the conserved residues are used to align the digestion pattern, it is clear that the digestion patterns of the two operators are not identical. One is forced to conclude that the contacts between the repressor and the two operators are not the same, and that the sequence differences in the operator sites cause subtle changes in the contact points between the DNA and the protein. Possibly this provides a mechanism for differential binding affinity of the repressor and its sites of regulation (*39*).

Although these studies demonstrate that it is feasible to convert a DNA binding protein into a site-specific nuclease, it is not possible to predict the likelihood of transforming any given DNA-binding protein into a nucleolytic agent. The generation of a site-specific nuclease by the method outlined requires that the derivatized lysine residue be unimportant for high affinity binding, yet proximal to the deoxyribose backbone in order for cleavage to proceed. However, these studies underscore the competence of the 1:1 1,10-phenanthroline-copper complex to generate the oxidative species capable of DNA scission.

Conclusions

The discovery of the nuclease activity of 1,10-phenanthroline-copper demonstrated that DNA could be readily cleaved by a coordination complex which binds to nucleic acid and generates reactive oxidative species. Other reactive groups, particularly ferrous EDTA [initially introduced by Hertzberg and Dervan (3)], can also degrade DNA efficiently when attached to appropriate ligands. In addition to providing new approaches for studying DNA conformational heterogeneity and protein-DNA interactions, these new reagents promise to provide an alternative to naturally occurring restriction enzymes in dissecting high molecular weight DNA's.

Literature Cited

1. *Nucleases*; Linn, S. M.; Roberts, R., Eds.; Cold Spring Harbor Laboratory, 1985.

2. Sigman, D. S.; Graham, D. R.; D'Aurora, V.; Stern, A. M. *J. Biol. Chem.* **1979**, *254*, 12269–12272.

3. Hertzberg, R. P.; Dervan, P. B. *J. Am. Chem. Soc.* **1982**, *104*, 313–315.

4. Tullius, T. D.; Dombroski, B. A. *Proc. Natl. Acad. Sci. USA* **1986**, *83*, 5469–5473.

5. Le Doan, T.; Perrouault, L.; Helene, C.; Chassignol, M.; Thuong, N. T. *Biochemistry* **1986**, *25*, 6736–6739.

6. Wood, B.; Skorobogaty, A.; Dabrowiak, J. C. *Biochemistry* **1987**, *26*, 6875–6883.

7. Barton, J. K. *Science* **1986** *223*, 727–734.

8. Hertzberg, R. P.; Dervan, P. B. *Biochemistry* **1984** *23*, 3934–3945.

9. Henner, W. D.; Rodriguez, L. O.; Hecht, S. M.; Haseltine, W. A. *J. Biol. Chem.* **1983**, *258*, 711–713.

10. Groves, J. T.; Subramanian, D. V. *J. Am. Chem. Soc.* **1984**, *106*, 2177.

11. Maxam, A.; Gilbert, W. *Meth. Enzymol.* **1980**, *65*, 499–599.

12. D'Aurora, V.; Stern, A. M.; Sigman, D. S. *Biochem. Biophys. Res. Comm.* **1977**, *78*, 170–176.

13. D'Aurora, V.; Stern, A. M.; Sigman, D. S. *Biochem. Biophys. Res. Comm.* **1978**, *80*, 1025–1032.

14. Marshall, L. E.; Graham, D. R.; Reich, K. A.; Sigman, D. S. *Biochemistry* **1981**, *20*, 244–250.

15. Pope, L. M.; Reich, K. A.; Graham, D. R.; Sigman, D. S. *J. Biol. Chem.* **1982**, *257*, 12121–12128.

16. Kuwabara, M.; Yoon, C.; Goyne, T. E.; Thederahn, T.; Sigman, D. S. *Biochemistry* **1986**, *25*, 7401–7408.

17. Goyne, T. E.; Sigman, D. S. *J. Amer. Chem. Soc.* **1987**, *109*, 2846–2848.

18. Dickerson, R. E.; Drew, H. R.; Conner, B. N.; Wing, R. M.; Fratini, A. V.; Kopka, M. L. *Science* **1982**, *216*, 475–485.

19. Kopka, M. L.; Yoon, C.; Goodsell, D.; Pjura, P.; Dickerson, R. E. *Proc. Natl. Acad. Sci. USA* **1985**, *82*, 1376–1380.

20. McClarin, J. A.; Frederick, C. A.; Wang, B.-C.; Greene, P.; Boyer, H. W.; Grable, J.; Rosenberg, J. M. *Science* **1986**, 1526–1541.

21. Pope, L. E.; Sigman, D. S. *Proc. Natl. Acad. Sci USA* **1984**, *81*, 3–7.

22. Yoon, C.; Kuwabara, M. D.; Law, R.; Wall, R.; Sigman, D. S. *J. Biol. Chem.* **1988**, *263*, 8458–8463.

23. Drew, H. R.; Travers, A. *Cell* **1984**, *37*, 491–502.

24. Law, R.; Kuwabara, M. D.; Briskin, M.; Fasel, M.; Hermanson, G.; Sigman, D. S.; Wall, R. *Proc. Natl. Acad. Sci. USA* **1987**, *84*, 9160–9164.

25. Sigman, D. S.; Spassky, A.; Rimsky, S.; Buc, H. *Biopolymers* **1985**, *24*, 183–197.

26. Spassky, A.; Sigman, D. S. *Biochemistry* **1985**, *24*, 8050–8056.

27. Gaynor, R. B.; Kuwabara, M. D.; Wu, F. K.; Garcia, J. A.; Harich, D.; Briskin, M.; Wall, R.; Sigman, D. S. *Proc. Natl. Acad Sci. USA* (in press).
28. Kuwabara, M.; Sigman, D. S. *Biochemistry* **1987**, *26*, 7234–7238.
29. Garner, M. M.; Revzin, A. *Nucleic Acids Res.* **1981**, *9*, 3047–3060.
30. Fried, M.; Crothers, D. M. *Nucleic Acids Res.* **1981**, *9*, 6505–6525.
31. Galas, D.; Schmitz, A. *Nucleic Acids Res.* **1978**, *5*, 3157–3170.
32. Spassky, A. *J. Mol. Biol.* **1986**, *188*, 99–103.
33. Chen, C.-H. B.; Sigman, D. S. *Proc. Natl. Acad. Sci. USA* **1986**, *83*, 7147–7151.
34. Chu, B.; Wahl, G.; Orgel, L. *Nucleic Acid Res.* **1983**, *11*, 6513–6529.
35. Inoue, T.; Cech, T. R. *Proc. Natl. Acad. Sci. USA* **1985**, *82*, 648–652.
36. *Organic Chemistry of Nucleic Acids, Part B*; Kochetkov, N. K.; Budovskii, E. I., Eds.; Plenum Press: London and New York, 1972; Ch. 8, pp 425–446.
37. Chen, C.-H. B.; Sigman, D. S. *Science* **1987**, *237*, 1197–1201.
38. Gunsalus, R. P.; Yanofsky, C. *Proc. Natl. Acad. Sci. USA* **1980**, *77*, 7117.
39. Kumamoto, A.; Arvidsen, D. N.; Gunsalus, R. *Genes Dev.* **1987**, *1*, 556.
40. Jue, R.; Lambert, J. M.; Pierce, L. R.; Traut R. R. *Biochemistry* **1978**, *17*, 5399.
41. Schevitz, R. W.; Otwinowski, Z.; Joachimiak, A.; Lawson, C. L.; Sigler, P. B. *Nature* **1985**, *317*, 782.
42. Kelley, R. L.; Yanofsky, C. *Proc. Natl. Acad. Sci. USA* **1985**, *82*, 483.

RECEIVED May 9, 1989

Chapter 3

Excited-State Modalities for Studying the Binding of Copper Phenanthrolines to DNA

Ramasamy Tamilarasan, David R. McMillin, and Fang Liu

Department of Chemistry, Purdue University, West Lafayette, IN 47907

Binding studies of copper(I) complexes of two phenanthrolines, dmp (2,9-dimethyl-1,10-phenanthroline) and bcp (2,9-dimethyl-4,7-diphenyl-1,10-phenanthroline), with DNA are described. The presence of the methyl substituents in the 2 and 9 positions allows us to focus on DNA binding because the reduction potential of the copper is too positive for significant nuclease activity. The results are consistent with simple surface association of Cu(dmp)$_2$+, presumably at the minor groove of DNA, and with intercalation of Cu(bcp)$_2$+ via the major groove. In the course of this investigation we have relied on the techniques of visible absorption spectroscopy and electrophoresis, as well as luminescence spectroscopy in the case of Cu(bcp)$_2$+. The luminescence results show that DNA forms a relatively rigid adduct with the bcp complex because solvent-induced quenching of the charge-transfer excited state—normally a very efficient process—is suppressed. Although Cu(dmp)$_2$+ and Cu(bcp)$_2$+ show negligible nucleolytic activity, these studies are relevant to the binding interactions of simple phenanthroline complexes that have been found to function as efficient, broad-spectrum artificial nucleases and useful probes of DNA structures (Sigman, D. S. Acc. Chem. Res. 1986, 19, 180–186).

IN 1979 SIGMAN AND CO-WORKERS REPORTED that a mixture of 1,10-phenanthroline (phen), copper(II), and thiol depolymerizes poly[d(A-T)] solutions under aerobic conditions (1). Subsequent work has demonstrated that copper phenanthrolines act as efficient, broad-spectrum artificial nucleases which are useful probes of DNA and DNA complexes (2–4). Hydrogen peroxide is an intermediate in the nuclease reaction since

0097–6156/89/0402–0048$06.00/0

catalase inhibits cleavage (*1*). A minimal reaction sequence is presented in Scheme I, where Red. and Ox. denote the reduced and the oxidized forms

Scheme I

$$Cu(NN)_2^{2+} + Red. \rightarrow Cu(NN)_2^+ + Ox. \tag{1}$$

$$2\ Cu(NN)_2^+ + O_2 + 2\ H^+ \rightarrow 2\ Cu(NN)_2^{2+} + H_2O_2 \tag{2}$$

$$Cu(NN)_2^+ + DNA \rightleftharpoons Cu(NN)_2^+ \bullet DNA \tag{3}$$

$$Cu(NN)_2^+ \bullet DNA + H_2O_2 \rightarrow oligonucleotides + Cu(NN)_2^{2+} + OH^- \tag{4}$$

of a sacrificial reagent such as a thiol, and $Cu(NN)_2^+ \bullet DNA$ represents a non-covalently bound adduct between the phenanthroline complex, $Cu(NN)_2^+$, and DNA. The binding of $Cu(phen)_2^+$ to DNA (Equation 3) is invoked to explain the high efficiency of the reaction and the sensitivity to substituents on the phen ligand (*3*). Reaction of the bound $Cu(phen)_2^+$ with hydrogen peroxide (Equation 4) appears to proceed via a hydroxyl radical-like species (*3*), presumably involving the cupryl moiety $Cu(OH)^{2+}$ (*5*), which attacks ribose linkages along the DNA polymer. The existence of a copper–oxygen bond in the intermediate is consistent with the fact that H_2O_2 is relatively inert as an outer-sphere oxidant (*6*) and, as discussed in more detail below, is entirely compatible with the coordination chemistry of the $Cu(phen)_2^{2+/+}$ couple. Electron transfer from copper is likely to occur after peroxide binds to copper although association and electron transfer could conceivably proceed in a concerted fashion.

Note that the putative cupryl intermediate is likely to be short-lived and perhaps unable to achieve an equilibrium distribution along the DNA polymer. If so, the binding of $Cu(NN)_2^+$ determines the reaction. We have begun a series of experiments designed to characterize the important binding event described by Equation 3. After a brief review of the ground- and excited-state chemistry of copper(I) phenanthroline complexes, we will describe how spectroscopic methods can be exploited to gain insight into the DNA binding phenomenon.

Ground- and Excited-State Chemistry

The key to isolating the binding equilibrium of $Cu(NN)_2^+$ with DNA is to select a phenanthroline derivative which forms a copper complex with a relatively positive reduction potential, thereby disfavoring Equation 2. This can be achieved with the introduction of methyl substituents at the 2 and 9 positions of the ligand (see Figure 1). Thus, the $Cu(dmp)_2^{2+/+}$ couple, where dmp denotes 2,9-dimethyl-1,10-phenanthroline, has a potential of 0.603 V vs. SHE compared to a potential of 0.035 V for the $Cu(phen)_2^{2+/+}$ couple (*7*, *8*). In fact, for the dmp system the reverse reaction, i.e., oxidation of H_2O_2 (or rather HO_2^-) by $Cu(dmp)_2^{2+}$, has been investigated (9). Since copper(I) is a d^{10} system–quite suited to sp^3 hybrid-ization–a four-coordinate, pseudotetrahedral coordination geometry is very common. In contrast, because of the half-vacant d orbital, four-coordinate copper(II) complexes tend to exhibit a flattened tetrahedral or even square planar geometry. To be sure, steric interactions between the 2,9-methyl substituents in $Cu(dmp)_2^{2+}$ destabilize the tetragonally dis-torted forms, but the main effect of the increased interligand repulsions may be to raise the energies of species with higher coordination numbers. Thus, $Cu(phen)_2^+$ (10) and $Cu(dmp)_2^+$ (11) are observed to have pseudo-tetrahedral coordination geometries in the solid state, whereas the copper(II) derivatives are found to be five-coordinate (12, 13), and even six-coordinate in the case of $Cu(phen)_2(NCS)_2$ (14). EXAFS methods have also been used to demonstrate that $Cu(dmp)_2^+$ picks up an extra ligand in solution upon oxidation (15). In view of the relatively large radius of the ion, it comes as no surprise that five-coordinate copper(I) complexes are also known (16, 17). Consequently, it would be possible for $Cu(phen)_2^+$ to reduce H_2O_2 by an inner-sphere mechanism which, as noted above, may be the kinetically favored path for both the oxidant and the reductant.

The change in ligation which occurs with an increase in the oxidation state of the copper ion is also a prominent factor shaping the excited state chemistry. Although copper(I) complexes do not exhibit ligand field transitions, phenanthroline complexes tend to be intensely colored due to low-lying metal-to-ligand charge transfer (CT) excited states. Excitation involves the transfer of an electron from a largely metal-based orbital to one which is composed mostly of a π^* antibonding molecular orbital of the ligand framework. Because of the vectorial nature of the charge displacement, such transitions can be strongly allowed, and they provide a sensitive handle on the complex in solution. Unless the non-radiative decay channels are too efficient, what is effectively the reverse process–CT luminescence–also serves as a useful probe of the solution environment (18, 19). At least in a formal sense the CT excited state can be regarded as a copper(II) center, and, not surprisingly, in the excited state the metal ion exhibits a tendency to add an additional ligand, e.g., a solvent molecule. The reaction provides a potent mechanism for coupling electronic and nuclear degrees of freedom and thereby facilitates electronic relaxation (19, 20). One of the consequences of this 'induced relaxation', or quench-ing, is a diminution of the emission intensity, which in turn serves as an experimental measure of the reaction efficiency.

If a new ligand-to-metal bond is formed in the excited state, consider-able reorganization of the existing ligand framework is required, as the basic coordination geometry changes from pseudotetrahedral to, for example, trigonal bipyramidal. At the outset of the present work we reasoned that intimate association with a large molecule like DNA could affect solvent-induced quenching by introducing a (further) barrier to the required structural reorganization. In the following we describe how absorption and emission spectroscopies have been used to study DNA binding.

Interaction with Cu(dmp)$_2$+

Figure 2 depicts the visible absorption spectrum of Cu(dmp)$_2$+ in 0.025 M Tris buffer (pH 7.8) in the presence and absence of salmon testes DNA, where DNA-P/Cu denotes the DNA phosphate- or nucleotide-to-copper ratio. The decrease in the absorption intensity and the red shift in the absorption maximum are classic indications of DNA binding which have been seen in studies involving other chromophores (21, 22). Using the method of LePecq and Paoletti (23) we have obtained additional evidence for the binding of Cu(dmp)$_2$+ to DNA in 0.025 M Tris buffer (pH 7.8). In the experiments pertaining to Figure 3, the DNA-P concentration was 1.02 $\times$ 10^{-4} M, and, for each copper concentration, the ethidium bromide (EB) concentration varied from 11.2 $\times$ 10^{-6} M to 51.5 $\times$ 10^{-6} M. The Scatchard plots are based on the EB fluorescence intensity which has been corrected for the inner-filter effect due to absorption by Cu(dmp)$_2$+. Since these corrections can be large and are difficult to make, the numbers obtained are only approximate values. The analysis yields an association constant of 5.0 $\times$ 10^6 M^{-1} for the binding of EB to DNA and 1.4 $\times$ 10^6 M^{-1} for the binding of the copper complex to DNA. In view of the smaller value of the ionic strength in our experiments the measured binding constant for EB is in reasonable agreement with previously reported values (23, 24). The data also suggest that the copper complex and EB bind competitively and that about 1.5 base pairs are required to define a binding site.

For Cu(dmp)$_2$+ only very weak CT emission can be observed in aqueous solution, and the emission intensity is not significantly affected by the addition of DNA. These results suggest that the bound form of Cu(dmp)$_2$+ remains fully accessible to solvent. In view of the close relationship between the structures it is reasonable to expect that the binding of Cu(dmp)$_2$+ is similar to that of Cu(phen)$_2$+. Consistent with this view, Graham and Sigman have reported that Cu(dmp)$_2$+ inhibits the nuclease activity of the Cu(phen)$_2$+ system, although the rate profile is qualitatively different than that observed for the intercalator EB (25). The nature of the binding of Cu(phen)$_2$+ is not understood in detail, but both the regiospecificity of hydrogen atom abstraction from the ribose unit (3) and footprinting studies of a structurally characterized DNA-drug adduct (26) indicate that reaction occurs at the minor groove of DNA. The minor groove appears to be too narrow to permit intercalation, hence the binding is viewed as a contact between two three-dimensional surfaces where the combined effects of electrostatic and hydrophobic forces, conditioned by short range interatomic repulsive interactions, determine the most favorable distribution and orientation of the copper complex along the groove.

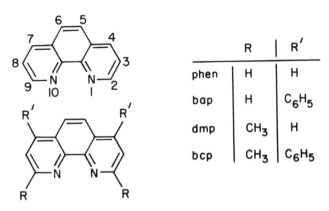

Figure 1. Ligands and ligand abbreviations.

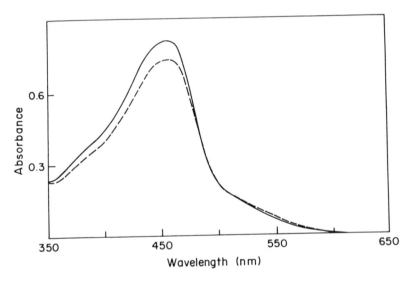

Figure 2. Absorption spectra of a 100 µM solution of Cu(dmp)$_2^+$ in 0.025 M Tris•Cl (pH 7.8) at 20 °C. With DNA–P/Cu = 0 (———); with DNA–P/Cu = 20 (– – –).

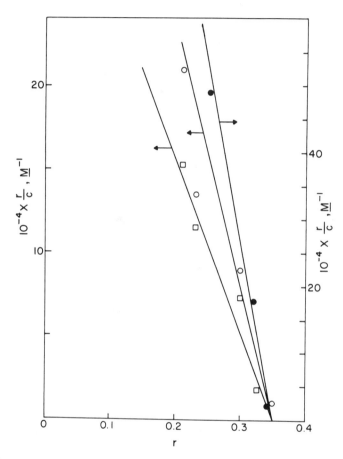

Figure 3. Scatchard plots derived from the ethidium bromide ion (EB) emission intensity. r represents the ratio of bound EB per DNA–P, and c represents the concentration of free EB. The DNA–P/Cu ratios are, from right to left, ∞ (●), 6 (○), and 4 (□).

Interaction with Cu(bcp)$_2$$^+$

With the phenyl substituents in the 4,7 positions Cu(bcp)$_2$$^+$ compounds are less soluble in water, and water/alcohol mixtures must be employed. (bcp denotes 2,9-dimethyl-4,7-diphenyl-1,10-phenanthroline.) When the percent by volume of methanol (MeOH) exceeds about 30%, [Cu(bcp)$_2$]Cl dissolves in solution; however, it is only sparingly soluble at lower alcohol levels. Kinetically stable solutions with an absorbance of 0.2–0.5 at the CT band maximum can be obtained in 6–20% MeOH by diluting a methanolic solution of the complex into aqueous buffer, but a detailed analysis suggests that these solutions actually represent colloidal suspensions (27).

When excess DNA is added to a solution containing Cu(bcp)$_2$$^+$ in 33% MeOH, a hypochromic effect is observed, indicative of DNA binding. The CT absorption maximum shifts from 474 nm to 478 nm and the absorbance decreases by about 7%. Here, however, excess DNA also has a striking effect on the emission spectrum in that it induces a greater than ten-fold increase in the quantum yield of CT emission (Figure 4). The DNA concentration dependence of the emission is, however, complex. As the DNA-P/Cu ratio is increased, the emission intensity initially increases and then maximizes at DNA-P/Cu ~ 2 (Figure 5). The emission intensity then decreases until a ratio of *ca.* 5, at which point the emission intensity starts to increase again until it plateaus at DNA-P/Cu > 20 (27). We have assigned the emission at high DNA-P/Cu ratios to copper monomers bound to DNA (27), and we have tentatively ascribed the emission at low DNA-P/Cu ratios to aggregated forms of the copper complex which are stabilized by contact with DNA. Similar concentration-dependent binding has been observed in studies involving relatively hydrophobic intercalating agents such as proflavin (21).

Obviously, the bcp complex binds to DNA in a different way than the dmp complex since there is no suppression of solvent-induced quenching for the dmp complex. This, and the fact that the CT emission from Cu(bcp)$_2$$^+$ is strongly polarized in the presence of DNA (27), can be explained if we assume that the bcp complex becomes rigidly entangled with the DNA molecule such that the copper complex cannot freely rotate or readily expand its coordination number. This type of binding is more consistent with intercalation than with the simple surface association proposed for Cu(dmp)$_2$$^+$. It should be noted in this regard that Barton and co-workers have previously proposed that 4,7-diphenyl substitution disposes phenanthroline complexes to intercalative binding *via* the major groove of DNA (28). There is a problem with this mode of interaction in that hydrogen atom/hydrogen atom repulsive interactions destabilize the conformation with the phenyl substituents in the plane of the phenanthroline moiety. Thus, in the solid state the phenyl group of the free ligand is well out of the plane, defining a dihedral angle of 63.2° (29). The difficulty in forming the planar conformation has been cited as the

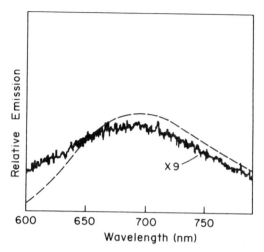

Figure 4. Emission spectra of 25 μM solutions of Cu(bcp)$_2^+$. In MeOH (——— ; × 9); in 33% MeOH:H$_2$O (V/V) containing 50 equivalents of DNA (– – –). The temperature was 20 °C, and the aqueous solution was nominally 0.025 M Tris•Cl buffer (pH 7.8).

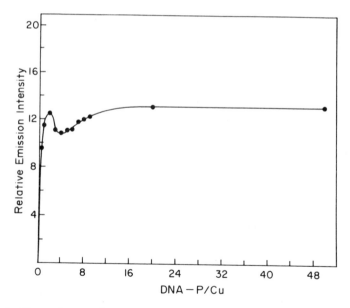

Figure 5. Dependence of CT emission intensity from a 25 μM solution of Cu(bcp)$_2^+$ on the DNA–P/Cu ratio. The solvent was 33% MeOH:H$_2$O (V/V), with 0.025 M Tris•Cl buffer (pH 7.8), and the temperature was 20 °C.

explanation of why N-methyl-4,7-diphenyl-1,10-phenanthrolinium ion does not intercalate into DNA whereas a number of other phenanthrolinium ions do (30). Even so, intercalative binding of Cu(bcp)$_2^+$ could still be possible if a non-planar conformation of the ligand is operative or if the binding energy is enough to overcome the barrier against rotation of the phenyl group into the plane. We recently obtained strong evidence for intercalative binding when we observed that Cu(bcp)$_2^+$, but not Cu(dmp)$_2^+$, retards the electrophoretic mobility of ϕX174 DNA when it is in the covalently closed supercoiled form (31, 32).

Summary and Outlook

We have shown that the binding of bis-phenanthroline complexes of copper(I) with DNA can conveniently be studied by utilizing phenanthroline derivatives with methyl substituents in the 2 and 9 positions. With these ligands the reduction potential of the Cu(II)/Cu(I) couple is sufficiently positive that there is no significant nuclease activity under normal assay conditions. In the case of Cu(dmp)$_2^+$ binding is evident from the hypochromic effect in the visible spectrum. Although Cu(dmp)$_2^+$ competes with ethidium for DNA, this particular copper complex does not intercalate into DNA, and all of the experimental results can be explained assuming a simple surface-to-surface contact with DNA. Quite different behavior is seen with Cu(bcp)$_2^+$ in a methanol/water solvent system. At high DNA-P/Cu ratios the bcp complex forms a relatively rigid adduct with DNA such that the efficiency of solvent-induced quenching of the CT excited state is significantly decreased. At the same time the CT emission is strongly polarized, and interaction with the bcp complex preferentially lowers the electrophoretic mobility of covalently closed supercoiled DNA. All of these results indicate that Cu(bcp)$_2^+$ binds by intercalation, presumably via the major groove of DNA. Sigman and co-workers have previously reported that Cu(phen)$_2^+$ and Cu(bap)$_2^+$ exhibit significantly different nucleolytic activity (3). These results are also readily accounted for if the presence of the phenyl substituents in the 4 and 7 positions of bap redirects the binding.

In contrast to the effect on the ground state, the presence of methyl substituents dramatically enhances the excited state redox activity of the copper systems by prolonging the lifetime (18, 19, 33). Future efforts will be directed in part towards the exploration of these complexes as photochemical probes of DNA.

Acknowledgments

This work has been supported by grants from the National Institutes of Health (GM 22764) and the National Science Foundation (CHE 87-19538).

Literature Cited

1. Sigman, D. S.; Graham, D. R.; D'Aurora, V.; Stern, A. M. *J. Biol. Chem.* **1979**, *254*, 12269–12272.
2. Spassky, A.; Sigman, D. S. *Biochemistry* **1985**, *24*, 8050–8056.
3. Sigman, D. S. *Acc. Chem. Res.* **1986**, *19*, 180–186.
4. Veal, J. M.; Rill, R. L. *Biochemistry* **1988**, *27*, 1822–1827.
5. Johnson, G. R. A.; Nazhat, N. B. *J. Am. Chem. Soc.* **1987**, *27*, 87–106.
6. Vu, D. T.; Stanbury, D. M. *Inorg. Chem.* **1987**, *26*, 1732–1736.
7. Hawkins, C. J.; Perrin, D. D. *J. Chem. Soc.* **1962**, 1351–1357.
8. Hawkins, C. J.; Perrin, D. D. *J. Chem. Soc.* **1963**, 2996–3002.
9. Davies, G.; Higgins, R.; Loose, D. J. *Inorg. Chem.* **1976**, *15*, 700–703.
10. Healy, P. C.; Engelhardt, L. M.; V. A.; White, A. H. *J. Chem. Soc. Dalton Trans.* **1985**, 2541–2545.
11. Dobson, J. F.; Green, B. E.; Healy, P. C.; Kennard, C. H.; Pakawatchai, C.; White, A. H. *Aust. J. Chem.* **1984**, *37*, 649–659.
12. Addison, A. W.; Palaniandavar, M.; Sinn, E. *Abstracts of Papers; 188th National Meeting of the American Society, Philadelphia, PA,* American Chemical Society: Washington, DC, **1984**; INOR 258.
13. Nakai, H.; Noda, Y. *Bull. Chem. Soc. Jpn.* **1978**, *51*, 1386–1390.
14. Sedov, A.; Dunaj–Juro, M.; Kabesôva, M.; Gazo, J.; Garaj, J. *Inorg. Chim. Acta* **1982**, *64*, L257–L258.
15. Elder, R. C.; Lunte, C. E.; Rahman, A. F. M. M.; Kirchhoff, J. R.; Dewald, H. D.; Heineman, W. R. *J. Electroanal. Chem.* **1988**, *240*, 361–364.
16. Gagne, R. R.; Allison, J. L.; Gall, R. S.; Koval, C. A. *J. Am. Chem. Soc.* **1977**, *99*, 7170–7178.
17. Goodwin, J. A; Stanbury, D. M.; Wilson, L. J.; Eigenbrot, C. W.; Scheidt, W. R. *J. Am. Chem. Soc.* **1987**, *109*, 2979–2991.
18. McMillin, D. R.; Gamache, R. E., Jr.; Kirchhoff, J. R.; Del Paggio, A. A. In *Biochemical and Inorganic Perspectives in Copper Coordination Chemistry*; Karlin, K. D., Ed.; Adenine: Guilderland, NY, **1983**; pp 223–235.
19. McMillin, D. R.; Kirchhoff, J. R.; Goodwin, K. V. *Coord. Chem. Rev.* **1985**, *64*, 83–92.
20. Palmer, C. E. A.; McMillin, D. R. *Inorg.Chem.* **1987**, *26*, 3837–3840.
21. Wilson, W. D.; Jones, R. L. In *Intercalation Chemistry*; Whittingham, M. S.; Jacobson, A. J., Eds.; Academic Press: New York, **1982**; 445–500.
22. Barton, J. K. *Science* **1986**, *233*, 727–734.
23. LePecq, J.–B.; Paoletti, C. *J. Mol. Biol.* **1967**, *27*, 87–106.
24. Jennette, K. W.; Lippard, S. J.; Vassiliades, G. A.; Bauer, W. R. *Proc. Natl. Acad. Sci. USA,* **1974**, *71*, 3839–3843.

25. Graham, D. R.; Sigman, D. S. *Inorg. Chem.* **1984**, *23*, 4188–4191.

26. Kuwarbara, M.; Yoon, C.; Goyne, T.; Thederahn, T.; Sigman, D. S. *Biochemistry* **1986**, *25*, 7401–7408.

27. Tamilarasan, R.; Ropartz, S.; McMillin, D. R., *Inorg. Chem.* **1988**, *27*, 4082–4084.

28. Barton, J. K.; Basile, L. A.; Danishefsky, A.; Alexandresen, A. *Proc. Natl. Acad. Sci. USA.* **1984**, *81*, 1961–1965.

29. Klemens, F. K.; Fanwick, P. E.; McMillin, D. R., submitted for publication.

30. Gabbay, E. J.; Scofield, R. E.; Barter, C. S. *J. Am Chem. Soc.* **1973**, *95*, 7850–7857.

31. Waring, M. *J. Mol. Biol.* **1970**, *54*, 247–279.

32. Tamilarasan, R.; McMillin, D. R., to be submitted.

33. Ichinaga, A. K.; Kirchhoff, J. R.; McMillin, D. R.; Dietrich-Buchecker, C. O.; Marnot, P. A.; Sauvage, J. P. *Inorg. Chem.* **1987**, *26*, 4290–4292.

RECEIVED March 3, 1989

Chapter 4

Interaction of Porphyrins and Metalloporphyrins with Nucleic Acids

Robert F. Pasternack[1] and Esther J. Gibbs[2]

[1]Department of Chemistry, Swarthmore College, Swarthmore, PA 19081
[2]Department of Chemistry, Goucher College, Towson, MD 21204

Following the discovery that the water soluble, cationic porphyrin tetrakis(4-N-methylpyridyl)porphine (H_2TMpyP-4) can intercalate into calf thymus DNA, studies in our laboratory on the interaction of a variety of its metal derivatives with synthetic nucleic acids revealed a correlation between the identity of the inserted metal and the mode and specificity of binding. A variety of spectroscopic signatures, and in particular induced circular dichroism spectra in the Soret region, have proved useful in distinguishing between intercalative and external binding. Kinetic results corroborate conclusions based upon these spectral patterns. The binding of porphyrins and metalloporphyrins to natural DNAs has been probed using similar methods and the models based upon work with synthetic polynucleotides seem largely to be applicable. Porphyrin derivatives capable of intercalation have been shown to be quite effective at converting Z- to B-form DNA. A mechanism has been proposed which involves the formation of a complex between the porphyrin and Z-DNA as an intermediate prior to the conformational changes. Metal derivatives containing axial ligands which have been shown not to intercalate promote this conversion much less efficiently, if at all.

INVESTIGATIONS OF THE INTERACTIONS of cationic *meso*-substituted porphyrins (*cf.* Figure 1) and their metalloderivatives with nucleic acids are proving useful in probing nucleic acid structure and dynamics. These studies are providing information fundamental to a more detailed understanding of the molecular basis of a variety of genetic processes and drug-nucleic acid phenomena in general. There are several advantages to

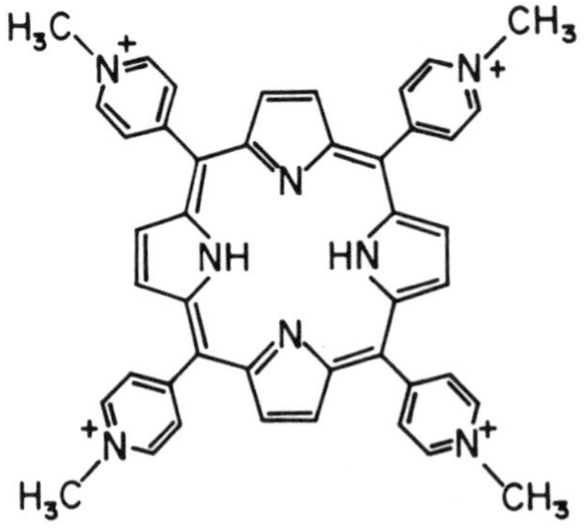

H₂TMpyP-4⁴⁺

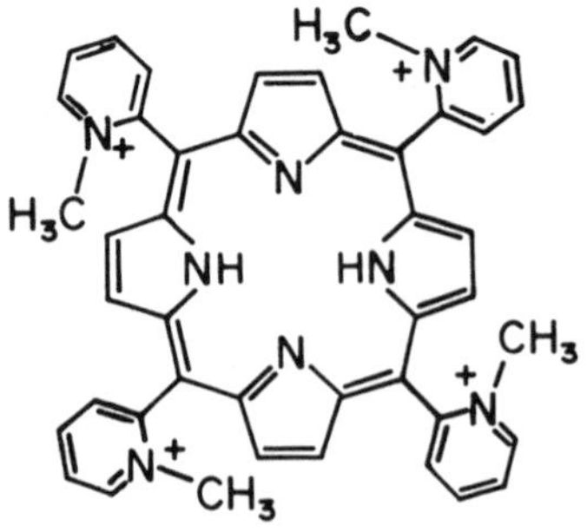

H₂TMpyP-2⁴⁺

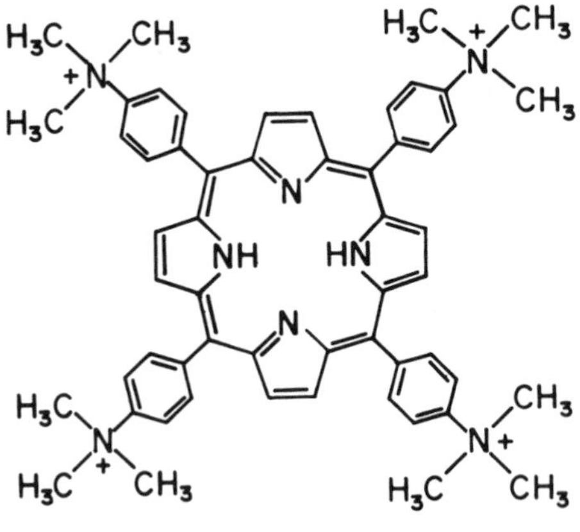

H₂TAP⁴⁺

Figure 1. Structures of several cationic water soluble porphyrins whose interactions with nucleic acids have been investigated.

utilizing these synthetic cationic porphyrins and metalloporphyrins as model drugs and probes. These species are quite water soluble and their solution chemistry has been extensively studied. They are intense chromophores and many fluoresce, providing convenient signals for monitoring reactions. The majority of these cationic porphyrins do not aggregate in the ground state, simplifying analysis. Lastly, modification of the structure of these porphyrins can be made by relatively simple synthetic procedures permitting systematic studies of the nature of the probe's interaction with the nucleic acid. These changes can, for example, alter the steric requirements of the porphyrin, modulate redox properties of the metal center and/or ring system, and even promote aggregation.

Studies with Synthetic Homo(purine-pyrimidine)$_2$ DNA's

Evidence that $H_2TMpyP-4$ can intercalate into calf thymus DNA was first presented by Fiel and co-workers (1, 2). This result was received with some surprise and even skepticism given the structural features of this molecule: X-ray crystal structures show the plane of the N-methylpyridyl groups to be at a 66–72° angle with respect to the porphine core (3) and NMR studies indicate that the rotational barrier for these N-methyl-pyridyl groups to achieve coplanarity with the porphine core is relatively large (4, 5). Thus considerable steric constraints would appear to exist for passage of these molecules between the base pairs of the DNA helix, as required to form an intercalated complex. Since the original report indicating porphyrin intercalation appeared, a number of studies utilizing a wide range of techniques have been conducted and a detailed picture of the mode(s) and mechanism(s) of this binding process has emerged (1, 2, 6–22).

Studies in our laboratory on the interaction of a variety of metal derivatives of $H_2TMpyP-4$ with synthetic nucleic acids revealed an interesting correlation between the identity of the inserted metal and the mode and specificity of binding to synthetic DNAs (6, 7). On the basis of visible absorption, circular dichroism spectroscopy and kinetic studies we suggested that $H_2TMpyP-4$ and its metal derivatives not having axial ligands [Au(III), Cu(II), Ni(II)] when interacting with poly(dG-dC)$_2$ are capable of intercalation (6, 7, 17). This intercalation process is accompanied by a *large* bathochromic shift and substantial hypochromicity of the Soret band, a *negative* induced CD band in the Soret region of the porphyrin chromophore (*cf* Table I) and binding kinetics that can be measured by temperature jump relaxation techniques. These same non-axially liganded porphyrins do not intercalate into poly(dA-dT)$_2$ but bind externally, probably in a groove of this polymer with some porphyrin-base overlap likely (3, 6, 7). The external complexes formed with poly(dA-dT)$_2$ display moderate bathochromic shifts and hypochromicity (in some cases a hyperchromicity) and a *positive* induced CD band in the Soret region of the porphyrin chromophore.

Table I. Changes in Porphyrin and Metalloporphyrin Soret Absorption and CD Maxima in the Presence of Nucleic Acids

Porphyrin	poly(dA-dT)$_2$ [a]		poly(dG-dC)$_2$ [a]		poly(dA-dC)·poly(dT-dG) [b]		Calf Thymus DNA [a]	
	Soret Δλ$_{max}$/%H	CD (nm)	Soret Δλ$_{max}$/%H	CD (nm)	Soret Δλ$_{max}$/%H	CD (nm)	Soret Δλ$_{max}$/%H	CD (nm)
H$_2$TMpyP-4	7/7	+433	21/41	−448	18/39	−440	7/40	+426 −448
Cu(II)TMpyP-4	3/−2	+419 −438(sm)	16/35	−439	10/28	−425 +415(sm)	6/18	−432 +415(sm)
Ni(II)TMpyP-4	7/20	+420 −439	17/46	−434	-----	-----	17/38	−437
Au(III)TMpyP-4	3.5/5	+401 −415(sm)	8/38	−408 +425(sm)	7/46	−408	8/35	−410
Zn(II)TMpyP-4	2/−6	+428 +438	0/4	+440 +450	5/−7	+425 +445	2/−5	+433 +453
Fe(III)TMpyP-4	5/9	+440	0/−2	no band	-----	-----	6/12	+433
H$_2$TMpyP-2	1/−2	+420	0/−3	no band	1/−1	+420	0/0	+415

[a] μ = 0.2 M; pH = 6.8; $1/r_0 \approx 30$; 25 °C; data from (6,17).

[b] μ = 0.1 M; pH = 7.0; $1/r_0 \approx 15$; 25 °C; [polymer] = 60 μM; data from (17).

The appearance of induced circular dichroism bands is due usually either to the introduction of a chiral center in a molecule as it forms a complex and/or to the coupling of a transition moment to that of a chiral molecule. This latter mechanism is the one believed to be responsible for porphyrin-induced CD spectra in complexes with nucleic acids. In the present studies, emphasis is on the electric dipole allowed $\pi \to \pi^*$ transition of the Soret band—a transition whose moment is reported to be in the plane of the molecule (23). A recent theoretical study (24) proposes that a positive induced effect for a non-intercalated molecule implies that the direction of the transition moment is along the groove, a porphyrin orientation on the exterior of DNA somewhat different from previous suggestions (3, 6). The question of whether externally bound porphyrins are partially intercalated or not is thus still to be resolved. The Ni(II) derivative of H_2TMpyP-4 is unusual in that it shows a conservative induced CD spectrum in its Soret region when bound to poly(dA-dT)$_2$. This exciton type CD spectrum is likely due to the self-stacking of NiTMpyP-4 on the surface of this polymer.

In contrast, metalloporphyrins which maintain their axial ligands when binding to nucleic acids [Zn(II), Co(III), Fe(III) and Mn(III)] do not intercalate into either poly(dG-dC)$_2$ or poly(dA-dT)$_2$ but bind externally to both polymers. The inability of these metalloporphyrins to intercalate is probably due to the additional steric requirements imposed by their axial ligands. This external binding to these nucleic acids leads to relatively *small* bathochromic shifts and hyper- or hypochromicity of the porphyrin Soret band, a *positive* or *no* induced CD of the metalloporphyrin chromophore and kinetic processes which are not accessible to joule heating temperature jump relaxation techniques. A detailed analysis of the data suggests that more stable external complexes are formed with the AT polymer than with the GC polymer (6, 7).

Carvlin, Fiel and coworkers considered binding patterns for the positional isomers of H_2TMpyP-4 at conditions similar to those of the above studies, importantly at **low drug load** (i.e., $r_o = $ [porphyrin]$_o$/[nucleic acid base pairs]$_o$ is small) (16). Based upon unwinding studies of covalently closed supercoiled DNA (ccs DNA), the 3-N-methyl isomer (H_2TMpyP-3) was shown to be an intercalator (12); it exhibits the negative induced CD band in the Soret region with poly(dG-dC)$_2$ indicative of intercalation. However, H_2TMpyP-2, in which the N-methyl groups are in the *ortho* position, and H_2TAP (*meso*-tetrakis(N,N,N-trimethylanilinium)porphine), another cationic porphyrin, do not unwind ccs DNA (12). Since intercalation is generally agreed to result in unwinding, these latter two porphyrins are considered to be non-intercalators. As would be predicted for external binding, with poly(dG-dC)$_2$ (at low r_o) H_2TMpyP-2 and H_2TAP display positive induced CD bands. All three species (H_2TMpyP-3, H_2TMpyP-2 and H_2TAP) exhibit positive CD bands with poly(dA-dT)$_2$ as does H_2TMpyP-4, again indicative of "external" groove binding. Therefore, the *binding mode preference of cationic meso-*

substituted porphyrins can be modulated not only by changes in the coordinated metal but also by minor alterations in the porphyrin periphery.

The following generalizations emerge for the binding of these porphyrins and metalloporphyrins to the synthetic homopolymers at low values of r_0: (1) intercalation occurs with poly(dG-dC)$_2$ but not poly(dA-dT)$_2$, (2) intercalation with poly(dG-dC)$_2$ occurs only with the non-axially liganded TMpyP-3 and TMpyP-4 porphyrins and metalloporphyrins, (3) negative induced CD bands in the Soret region of these porphyrin chromophores are indicative of intercalation, (4) positive induced CD bands are a signature for "external" binding, and (5) poly(dA-dT)$_2$ forms more stable external complexes with these porphyrins and metalloporphyrins than does poly(dG-dC)$_2$.

The bonding mode specificity of these porphyrins with synthetic polymers is likely the result of thermodynamic and steric effects. The more stable poly(dG-dC)$_2$ helix (the melting temperature of GC polymers is much higher than that of AT polymers) is capable of holding the large porphyrin in its interior as an intercalated complex whereas the poly(dA-dT)$_2$ polymer may not be stable enough to accomplish this. Whereas intercalation of other molecules into poly(dA-dT)$_2$ has been reported, these drugs usually have smaller steric barriers to intercalation and/or are capable of simultaneous groove binding which helps stabilize the complex (25). Molecular modeling studies with H$_2$TMpyP-4 suggest that intercalation into TpA sites is sterically hindered due to the presence of the methyl group on thymine (3). It has also been suggested that the formation of an external complex in a groove of GC polymers may be partially blocked by the 2-amino group of guanine that protrudes into the minor groove (26, 27). In addition, the more flexible poly(dA-dT)$_2$ polymer is able to bend, flex or kink around molecules bound externally in its minor groove, increasing the number and strength of interactions between the drug and the polymer thereby stabilizing external complexes. Such a process is disfavored for the less flexible and more stable GC duplex. Outside binding of the polycationic porphyrin molecules at AT sites may also be enhanced by the higher electrostatic potential of AT polymers relative to GC polymers (28–30), resulting in an increase in the coulombic force of attraction.

That H$_2$TMpyP-2 is a non-intercalator may be due to the larger barrier to rotation of the *meso*-N-methylpyridyl rings when the N-methyl group is in the *ortho* position as compared to its being in the *meta* or *para* position; some degree of rotation of these pyridyl rings (not necessarily to coplanarity to the porphine core) may be required for passage of the porphyrin into the DNA helix or to achieve the proper orientation of the pyridyl rings relative to the DNA surface in the stable intercalated complex. Even this small degree of rotation may be energetically too unfavorable for the *ortho* isomer to allow intercalation. But other factors such as the distance between the positive charge of the N-methyl group and the negative phosphate group may also be important as well as steric problems arising from the position of the N-methyl group relative to the duplex. The role that each of these various factors is playing in the binding mode and base specificity of these porphyrins and metalloporphyrins is currently not completely understood but studies in our laboratory are proceeding to address this question.

Interactions with Mixed Base Pair Deoxyribonucleic Acids

At small values of ro (ro < 0.1), $H_2TMpyP-2$, H_2TAP and the *axially liganded* metalloporphyrins display positive induced CD bands with calf thymus DNA (*6, 15*). From the previously described work with the synthetic homopolymers this would suggest that these porphyrins and metalloporphyrins are binding externally to DNA. On the other hand, the *nonaxially liganded* metalloporphyrins have negative induced CD bands in their Soret region, suggesting that they are highly intercalation specific with calf thymus DNA (*6*). This assignment of binding mode with DNA is corroborated by absorbance, kinetic (*6, 7*), melting temperature (*1*), linear dichroism (*18*), flow dichroism (*10, 11, 21*), NMR (*11, 31*), ESR (*32*) viscosity (*10, 11*) and unwinding (*2, 12, 16, 19, 21*) studies. For example, ESR spectra of pressed films provide unequivocal evidence for the intercalation of CuTMpyP-4 into calf thymus DNA and those porphyrins which have positive CD bands with calf thymus DNA do not unwind ccs DNA whereas those porphyrins which have negative CD bands do cause unwinding. We have therefore suggested that the sign of the induced CD as a marker for **binding mode** can be extended to natural DNAs at *small values of r_0* ($r_0 < 0.1$). It is important though to recognize that the indicated binding mode preference of these porphyrins and metalloporphyrins with natural DNAs does not also imply base pair specificity. One technique which can be quite useful in addressing the question of base pair specificity of binding is DNA footprinting. In this technique gel electrophoresis is used to determine which sites on a DNA fragment of known sequence are protected by drug binding from enzyme cleavage. The footprinting experiments of Ward et al. (*33*) indicate that external binding for axially liganded metalloporphyrins is *strongly favored* in regions containing strings of AT base pairs. But the base pair specificity for intercalation is less certain. In footprinting experiments with intercalating porphyrins at moderate levels of drug load there is no change in the cleavage pattern of the enzyme. In addition, absorbance, circular dichroism and kinetic experiments indicate that $H_2TMpyP-4$ and its Cu(II) and Au(III) derivatives intercalate into poly(dA-dC)•poly(dT-dG) (*17*). Viscosity experiments also suggest the nonmetallo and Ni(II) porphyrins can intercalate into this mixed base pair polymer (*21*). Thus, contiguous GC base pairs are likely not required for intercalation. What percentage and/or sequence(s) of GC base pairs are required for these non-axially liganded metalloporphyrins to intercalate, as well as the basis for the sequence dependence, is at present uncertain. We will return to this point in a later section.

Evidence obtained in our laboratories on the interaction of the metal-free $H_2TMpyP-4$ porphyrin with calf thymus DNA, at *low values of r_0*, led us to suggest that this porphyrin shows intercalation specificity at low ionic strength and a preference for external binding at high ionic strength (8). At low salt concentrations we observed only a negative induced CD band. As the ionic strength is increased, the negative induced CD band decreases in intensity while a positive induced CD band begins to appear. The absorbance spectrum of $H_2TMpyP-4$ in the Soret region, at conditions in which both the positive and negative induced CD bands are present, appears not as a single peak but instead as a broad unsymmetrical band, indicating that there is more than one type of bound complex. This change in the nature of porphyrin binding mode with ionic strength is also suggested by unwinding studies of Carvlin (12), fluorescence studies of Kelly et al. (19) and linear dichroism work of Geacintov et al. (18) Another metal-free derivative, tris(4-N-methylpyridyl)monophenylporphine $(H_2(Mpy-4)_3\phi P)$ behaves much the same as does the parent compound $(H_2TMpyP-4)$ with, at most, some additional preference for outside bonding (Gibbs, Ellinas and Pasternack, Swarthmore College, unpublished data). This effect appears to be peculiar to the metal-free derivatives; none of CuTMpyP-4, NiTMpyP-4, AuTMpyP-4 nor ZnTMpyP-4 show a significant change in binding mode preference to DNA with ionic strength (at low value of r_0).

It is in considering the binding of the *non-metalloporphyrin* $H_2TMpyP-4$ to natural DNAs that several alternative models have been presented. From ^{31}P NMR studies of $H_2TMpyP-4$ with a number of oligomers and the poly(dG-dC)$_2$ polymer, Marzilli and his co-workers suggest that $H_2TMpyP-4$ intercalates selectively at 5'CpG3' sites (31). Using this premise of highly selective intercalation as a basis, these workers offer two models for binding of $H_2TMpyP-4$ to calf thymus DNA, both of which require "external" binding involving considerable stacking of $H_2TMpyP-4$ molecules, even at small values of r_0. This stacking interaction must be invoked to account for the large bathochromic shift and hypochromicity of the Soret absorbance band observed for the $H_2TMpyP-4 \bullet DNA$ complexes at low r_0 while at the same time limiting intercalation to 5'CpG3' sites. Although the ^{31}P NMR studies are suggestive of 5'CpG3' selectivity for $H_2TMpyP-4$ intercalation into oligomers, they are not definitive. In the presence of $H_2TMpyP-4$, a ^{31}P NMR peak at –1 ppm is observed with the 5'-d(TATACGCGTATA)-3' and poly(dG-dC)$_2$ duplexes but is not observed with the 5'-d(TATATGCATATA)$_2$-3' duplex. Marzilli et al. thus conclude that $H_2TMpyP-4$ does not intercalate into 5'GpC3' sites. However, it may well be that for oligomers there is a minimum number of GC base pairs required to form an intercalation site. If the stability of local helical regions is important for intercalation, longer strings of GC base pairs may be required for intercalation to occur, especially when they are located in predominantly AT rich regions of a nucleic acid. All the oligomers for which Marzilli et al. observed intercalation of $H_2TMpyP-4$ have strings of four GC base pairs, whereas all the oligomers in which ^{31}P NMR results indicated intercalation was not occurring have GC base pair strings of less than four. More recent NMR and viscosity studies suggest that other intercalating *metalloporphyrins* do not show this CpG specificity, so the effect if it occurs is peculiar to $H_2TMpyP-4$ (21). Yet Ford et al. observe in footprinting experiments (34) that $H_2TMpyP-4$ and its nickel derivative produce identical results, i.e., the metal ion plays *no role in sequence selectivity*. Additional NMR studies on oligomers of selected sequence, in conjunction with other spectroscopic techniques should be of enormous value in addressing the question of what constitutes an intercalation site for these porphyrins.

In footprinting experiments performed by Ward et al. (22), neither $H_2TMpyP-4$ nor its Ni(II) and Cu(II) derivatives had any influence on the cutting pattern of DNA up to an r_0 value of 0.2. Since with the *axially liganded metalloporphyrins* of TMpyP-4 (all of which bind only externally to DNA) there is protection of the restriction fragment from cleavage at r_0 values as low as 0.02, it would appear that $H_2TMpyP-4$ and the Cu(II) and Ni(II) derivatives are not binding extensively to the exterior of DNA at moderate levels of drug load. We believe that at present the sum of all the experimental evidence weighs heavily against the stacking model, particularly at low levels of drug load. Above $r_0=0.2$, Ward et al. observed nonselective protection of the restriction fragment from hydrolysis by intercalating porphyrins and metalloporphyrins. This could be the result of the conformational changes of the DNA caused by these intercalators in this drug load range and/or the beginning of external binding. In similar footprinting experiments, Ford and co-workers also observed GC and AT site protection by $H_2TMpyP-4$ and its Ni(II) derivative (34). On the basis of spectral (absorbance and CD) and kinetic experiments we suggested that at high *excess DNA* (low r_0) and moderate ionic strength the nickel and copper derivatives of TMpyP-4 show a stronger intercalation *binding mode preference* than the nonmetallo $H_2TMpyP-4$ (6, 7). This is not in contradiction to the suggestion by Marzilli et al. that $H_2TMpyP-4$ is more *site specific* in its intercalation binding than is NiTMpyP-4. But in fact it is not possible at this time to distinguish between greater intercalation site selectivity of $H_2TMpyP-4$ versus its increased ability to form stable external complexes with natural DNAs, as expected on the basis of ionic strength experiments discussed earlier.

At high levels of drug load ($r_0 > 0.2$) the binding of $H_2TMpyP-2$, -3 or -4 and H_2TAP porphyrins to nucleic acids becomes more complex and the resultant induced CDs more complicated. In some cases changes in the CD of the nucleic acid are also observed. These more complicated induced CD spectra of the porphyrins and the changes in the CD of the nucleic acid are due to the variety of complexes which can form at high concentrations of drug as a result of: (i) conformational changes of the polymers, (ii) saturation of intercalation sites resulting in binding of intercalating porphyrins on the exterior of poly(dG-dC)$_2$ or natural DNA polymers, (iii) stacking or aggregation of "externally" bound porphyrin molecules on the nucleic acid surface and/or (iv) aggregation of the polymers. Studies with porphyrins which have a greater tendency to aggregate and particularly with nucleic acids which promote stacking of porphyrins (i.e., polymers with a high percentage of AT sites) have complex CD features at lower r_0 values than observed for porphyrins which do not have a large tendency to aggregate. One example of the former type is the non-intercalating H$_2$TAP, studied extensively by Fiel and co-workers (15).

Kinetic Studies with B-form Nucleic Acids

Kinetic studies on the binding of H$_2$TMpyP-4 to poly(dG-dC)$_2$ were conducted using both joule (7) and laser heating (Gibbs, Holzwarth and Pasternack, Fritz Haber Institute, unpublished results) temperature jump relaxation techniques. Identical results were obtained for both heating methods in contrast to kinetic results obtained for other drug-nucleic acid interactions. Marcandalli et al. (35) have recently repeated studies on the kinetics of binding of a number of cationic intercalating drugs, e.g., proflavin and ethidium, with a variety of nucleic acids using laser heating rather than the joule heating relaxation technique. Unlike the previous work in which multiphasic kinetic profiles were obtained for drug binding, Marcandalli et al. observed only monophasic relaxation profiles using a laser heating method. Pressure-jump relaxation methods also give monophasic kinetic profiles for ethidium•poly(dG-dC)$_2$ (36) and proflavin•calf thymus DNA interactions (37). The multphasic relaxation profiles obtained in earlier work for the binding of intercalators to DNA is believed to be due to orientational processes caused by the joule heating technique (35). That we do not observe these effects with the porphyrin intercalation system may be due to the slower kinetics of binding of the bulky H$_2$TMpyP-4, putting the reactions in a time range slow enough as not to couple with orientation phenomena caused by the joule heating method.

For the interaction of poly(dG-dC)$_2$ with H$_2$TMpyP-4 we observed a single relaxation effect using either heating method whose relaxation time and amplitude show a concentration dependence (7). A plot of τ^{-1} (τ = relaxation time) versus "corrected" concentration gives a straight line over the entire concentration range studied. It is generally accepted that intercalation occurs through the following two step mechanism (36, 38):

$$\text{Porphyrin} + \text{NA} \overset{K_{ext}}{\Longleftrightarrow} \text{Porphyrin:NA}_{external} \qquad \text{fast} \qquad (1)$$

$$\text{Porphyrin:NA}_{external} \underset{k_{-2}}{\overset{k_2}{\Longleftrightarrow}} \text{Porphyrin:NA}_{intercalated} \qquad \text{slow} \qquad (2)$$

The relaxation time for this mechanism is given by:

$$1/\tau = \frac{k_2([\overline{NA}] - f'(r)[\overline{P}])}{1/K_{ext} + ([\overline{NA}] - f'(r)[\overline{P}])} + k_{-2} \tag{3}$$

where f'(r) is a statistical factor included to take into account neighbor exclusion phenomena of intercalation binding and [$\overline{NA}$] is the concentration of the nucleic acid under study expressed in moles of base pairs per liter.

If K_{ext}, the equilibrium constant for external binding, is small as has been previously suggested from spectroscopic studies with poly(dG-dC)$_2$, then the concentration term in the denominator of Equation 3 becomes insignificant and the observed linear dependence of $1/\tau$ versus ([$\overline{GC}$] − f'(r)[$\overline{P}$]) is predicted. Using $K_{ext}(1+K_{int}) = 7.7 \times 10^5$ M^{-1} obtained from titration data and the data from this kinetic study, we obtain $k_{-2} = 1.8$ s^{-1}, $K_{ext} < 3000$ M^{-1} and $K_{int}/K_{ext} > 250$ for H$_2$TMpyP-4 with poly(dG-dC)$_2$. Thus, the intercalated complex is indeed considerably more stable than the complexes which form on the exterior of this polymer.

With poly(dA-dT)$_2$ and H$_2$TMpyP-4 we observed no relaxation effects over a range of concentrations of reactants. From classical hand mixing and stopped-flow kinetic experiments we know that the half life for this binding process has to be shorter than several milliseconds and the process is therefore faster than binding to poly(dG-dC)$_2$ (7). The lack of relaxation effects indicates that either the binding of this porphyrin to poly(dA-dT)$_2$ is too fast for even joule heating temperature jump relaxation techniques (half life < 20 μs) and/or the enthalpy change for this reaction is near zero (8). Kinetic relaxation studies of other drugs, such as proflavin or ethidium bromide, which are known to intercalate between AT base pairs, display temperature jump relaxation effects with poly(dA-dT)$_2$ (36 and references therein). This very different kinetic profile for H$_2$TMpyP-4 with poly(dA-dT)$_2$ is in accord with the suggestion from spectroscopic studies that the type of complex formed with AT polymers is substantially different from the type formed with the GC duplex.

With calf thymus DNA, using either laser or joule heating methods, this same non-metalloporphyrin and all of the intercalating metalloporphyrins yield large multiphasic relaxation effects whose rate constants do not change with concentration of porphyrin or nucleic acid (7, 17). To our knowledge this is the only study of an intercalator which gives such a kinetic profile with DNA. The lack of concentration dependence of the rates can be explained if it is assumed that K_{ext} is large compared to ([$\overline{NA}$] − f'(r)[$\overline{P}$]) and therefore Equation 3 reduces to $1/\tau = k_2 + k_{-2}$. That

K_{ext} is large is not unreasonable due to the presence of the AT sites in calf thymus DNA. Relaxation kinetic studies with solutions containing mixtures of poly(dA-dT)$_2$ and poly(dG-dC)$_2$ reveal that under the conditions of these experiments (μ = 0.15 M) the external complexes formed on the AT polymer have slightly larger binding constants than for intercalation into the GC polymer (7). Similar results were obtained by Kelly et al. using fluorescence techniques (19). The multiphasic nature of the relaxation effects could be a result of there being more than one type of external and/or intercalated complex or due to changes in the conformation of the DNA molecule. Thus a model that emerges is one in which the porphyrin binds rapidly to the external sites of AT regions of the heteropolymer and then relocates into an intercalation site, without dissociating into the bulk solvent medium. In contrast, the transfer of $H_2TMpyP-4$ between poly(dG-dC)$_2$ and poly(dA-dT)$_2$ follows a pathway involving dissociation of the porphyrin moiety from the polymers (7).

Z→B Conversion

Spectroscopic and kinetic studies were conducted with these porphyrins and poly(dG-dC)$_2$ at (i) $Co(NH_3)_6^{3+}$ concentrations (9) and (ii) alcohol concentrations in which the polymer is in the left-handed Z-conformation (20). As found for other intercalating drugs the intercalating porphyrins and metalloporphyrins are capable of converting Z-poly(dG-dC)$_2$ to the right handed B-conformation. Kinetic studies imply that these porphyrin species intercalate into the Z-form prior to B formation although some uncertainty exists as to the nature of drug-Z-DNA interactions. A recent report (39) proposes that proflavin (and possibly ethidium) binds to the sugar-phosphate backbone of Z-DNA rather than between base pairs, although end effects proved significant with the tetranucleotides investigated. Our work with porphyrins indicates that the kinetic intermediate formed prior to conformational conversion has the spectral characteristics of an intercalated complex. Each bound porphyrin molecule then converts multiple base pairs from Z→B form, implying that the interaction influences the structure of the polymer several base pairs beyond the binding site. The non-intercalating metalloporphyrins studied were not nearly as effective at this conversion. $H_2TMpyP-2$ is quite unusual, in that although it is a nonintercalator it is as efficient as the intercalators in converting Z- to B-DNA, but only in the alcohol medium and only above a threshold value of $r_0 > 0.03$. It has been suggested that conversion of DNA between the Z- and B-forms could be one mechanism for control of genetic processes (40). If this is the case then these intercalating porphyrins could interfere with such control mechanisms by binding to Z-regions of DNA and converting them to the B-form or by preventing conversion of B-DNA to the Z conformation in vivo.

Summary

Differences in the interpretation of the details of the binding of porphyrins and metalloporphyrins to nucleic acids exist, in part arising from the fact that different experimental techniques require different experimental conditions. For example, uv/vis absorption and CD spectroscopy utilize micromolar and 10–100 μM concentrations of drug and polymer respectively, whereas NMR utilizes millimolar concentrations, increasing the likelihood of DNA aggregation. Differences in solvent conditions can have a profound effect on binding; changes in binding mode with changes in ionic strength have been observed for two of the porphyrins studied. At high levels of drug load the nucleic acid conformation may be altered and/or new modes of binding may become important. Thus, care must be taken when comparing data at different experimental conditions. At the same time that these various complexities for binding make comparisons difficult, it is precisely because of such complexities that these porphyrins are such sensitive probes of nucleic acid structure and dynamics. It may well be the diversity of binding of these porphyrins that will provide a variety of avenues for therapeutic strategies. It is certainly true that obtaining a complete binding picture for porphyrins requires extended studies using a variety of techniques and experimental conditions. In spite of these complexities, the following generalizations of the data appear possible: **(1) minor changes in porphyrin structure are observed to cause major changes in the nature of their interaction with nucleic acids, (2) H_2TMpyP-2 does not intercalate, (3) H_2TAP does not intercalate but binds on the exterior of nucleic acids with extensive self-aggregation, (4) H_2TMpyP-4 and its metal derivatives form a variety of types of complexes with nucleic acids: (i) at low values of r_0, the type of complex formed can be correlated with the nature of the metal and can be gleaned from the sign of the induced CD in the Soret region of the porphyrin, (ii) at higher values of r_0, most of these porphyrin molecules induce changes in the conformation and/or aggregation of DNA, (5) the nonmetallo H_2TMpyP-4 and $H_2(Mpy-4)_3\phi P$ porphyrins show a change in binding mode with ionic strength, (6) AT regions of nucleic acids form more intimate and stable external complexes than do GC regions of nucleic acids, and (7) the intercalating porphyrins and under certain (low salt) conditions H_2TMpyP-2 convert Z- form DNA to the B conformation.**

Acknowledgments

The authors wish to gratefully acknowledge financial support for this work from the National Institutes of Health (GM 34676), the National Science Foundation (CHE 8613592) and the Monsanto Corporation.

Literature Cited

1. Fiel, R. J.; Howard, J. C.; Mark, E. H.; Datta-Gupta, N. *Nucleic Acids Res.* **1979**, *6*, 3093.
2. Fiel, R. J.; Munson, B. R. *Nucleic Acids Res.* **1980**, *8*, 2835.
3. Ford, K.; Pearl, L. H.; Neidle, S. *Nucleic Acids Res.* **1987**, *15*, 6553.
4. Eaton, S. S.; Eaton, G. R. *J. Am. Chem. Soc.* **1975**, *97*, 3600.
5. Eaton, S. S.; Fiswild, D. M.; Eaton, G. R. *Inorg. Chem.* **1978**, *17*, 1542.
6. Pasternack, R. F.; Gibbs, E. J.; Villafranca, J. J. *Biochemistry* **1983**, *22*, 2406.
7. Pasternack, R. F.; Gibbs, E. J.; Villafranca, J. J. *Biochemistry* **1983**, *22*, 5409.
8. Pasternack, R. F.; Garrity, P.; Ehrlich, B.; Davis, C. B.; Gibbs, E. J.; Orloff, G.; Giartosio, A.; Turano, C. *Nucleic Acids Res.* **1986**, *14*, 5919.
9. Pasternack, R. F.; Sidney, D.; Hunt, P. A.; Snowden, E. A.; Gibbs, E. J. *Nucleic Acids Res.* **1986**, *14*, 3927.
10. Banville, D. L; Marzilli, L. G.; Strickland, J. A. *Biopolymers* **1986**, *25*, 1837.
11. Banville, D. L.; Marzilli, L. G.; Wilson, W. D. *Biochem. Biophys. Res. Comm.* **1983**, *113*, 148.
12. Carvlin, M. J. Ph.D. Thesis, SUNY, Buffalo, **1985**.
13. Carvlin, M. J.; Datta-Gupta, N.; Fiel, R. J. *Biochem. Biophys. Res. Comm.* **1982**, *108*, 66.
14. Carvlin, M.; Datta-Gupta, N., Mark, E. H.; Fiel, R. J. *Cancer Res.* **1981**, *4*, 3543.
15. Carvlin, M.; Fiel, R. J. *Nucleic Acids Res.* **1983**, *11*, 6121.
16. Carvlin, M.; Mark, E. H.; Fiel, R. J. *Nucleic Acids Res.* **1983**, *11*, 6141.
17. Gibbs, E. J.; Maurer, M. C.; Zhang, J. H.; Reiff, W. M; Hill, D. T.; Malicka-Blaszkiewicz, M.; McKinnie, R. E.; Liu, H.-Q.; Pasternack, R. F. *J. Inorg. Biochem.* **1988**, *32*, 39.
18. Geacintov, N. E.; Ibanez.; Rougee, M.; Benasson, R. V. *Biochemistry* **1987**, *28*, 3087.
19. Kelly, J. M.; Murphy, M. J.; McConnell, D. J.; Oh Uigin, C. *Nucleic Acids Res.* **1985**, *13*, 16.
20. McKinnie, R. E.; Choi, J. D.; Bell, J. W.; Gibbs, E. J.; Pasternack, R. F. *J. Inorg. Biochem.* **1988**, *32*, 207.
21. Strickland, J. A.; Banville, D. L.; Wilson, W. D.; Marzilli, L. G. *Inorg. Chem.* **1987**, *26*, 3398.
22. Ward, B.; Skorobogarty, A.; Dabrowiak, J. C. *Biochemistry* **1986**, *25*, 7827.
23. Gouterman, M. In *The Porphyrins*, Dolphin, D., Ed.; Academic: New York, **1978**; Vol. 3A, pp 87–100.
24. Kubista, M.; Akerman, B.; Norden, B. *J. Phys. Chem.* **1988**, *92*, 2352.

25. Searle, M. S.; Hall, J. G.; Denny, W. A.; Wakelin, L. P. G. *Biochemistry* **1988**, *27*, 4340.
26. Kopka, M. L.; Yoon, C.; Goodsell, D.; Pjura, P.; Dickerson, R. E. *Proc. Natl. Acad. Sci. USA* **1985**, *82*, 1376.
27. Kopka, M. L.; Yoon, C.; Goodsell, D.; Pjura, P.; Dickerson, R. E. *J. Mol. Biol.* **1985**, *183*, 553.
28. Lavery, R.; Pullman, B. *J. Biomol. Struct. Dyn.* **1985**, *2*, 1021.
29. Pullman, B. *J. Biomol. Struct. Dyn.* **1983**, *1*, 773.
30. Weiner, P. K.; Langridge, R.; Blaney, J. M.; Schaefer, R.; Kollman, P. A. *Proc. Natl. Acad. Sci. USA* **1982**, *79*, 3754.
31. Marzilli, L. G.; Banville, D. L.; Zan, G.; Wilson, W. D. *J. Am. Chem. Soc.* **1986**, *108*, 4188.
32. Dougherty, G.; Pilbrow, J. R.; Skorobogaty, A; Smith, T. D. *J. Chem. Soc. Faraday Trans.* 2 **1985**, *81*, 1739.
33. Ward, B.; Skorobogaty, A.; Dabrowiak, J. C. *Biochemistry* **1986**, *25*, 6875.
34. Ford, K.; Fox, K. R.; Neidle, S.; Waring, M. J., *Nucleic Acids Res.* **1987**, *15*, 2221.
35. Marcandalli, B.; Winzek, C.; Holzwarth, J. F. *Bunsenges. Phys. Chem.* **1984**, *88*, 368.
36. Macgregor, R. B.; Clegg, R. M.; Jovin, T. M. *Biochemistry* **1987**, *26*, 4008.
37. Marcandalli, B.; Knoche, W.; Holzwarth, J. *Gazz. Chim. It.* **1986**, *116*, 417.
38. Crothers, D. M. *Prog. Mol. Subcell,. Biol.* **1970**, *2*, 11.
39. Westhof, E.; Hosur, M. V.; Sundaralingam, M. *Biochemistry* **1988**, *27*, 5742.
40. Rich, A.; Nordheim, A.; Wang, A. H.-J. *Ann. Rev. Biochem.* **1984**, *53*, 791.

RECEIVED June 13, 1989

Chapter 5

Porphyrins as Probes of DNA Structure and Drug–DNA Interactions

G. Raner, J. Goodisman, and J. C. Dabrowiak

Department of Chemistry, Syracuse University, Syracuse, NY 13244–1200

The water-soluble manganese porphyrin complex MnT4MPyP is versatile as a probe of local DNA structure and as an agent for studying drug-DNA interactions. The compound possesses a high specificity for trinucleotide sites having only adenine and thymine, i.e. (A•T)$_3$. Optical studies and the DNA strand scission pattern of the porphyrin on a 139-base pair restriction fragment of pBR322 DNA indicate that binding occurs via the minor groove and that local DNA melting is important in the binding mechanism. The metal complex has also been used in footprinting experiments involving the antiviral agent netropsin. Calculation of binding constants from the footprinting data requires a detailed analysis of the netropsin and porphyrin site exclusions on DNA. Due to the site-specific cleavage pattern of the porphyrin, its reporting of site occupancy by drug is not the same as that of DNase I. This factor must be addressed in developing the theory for quantitative studies with MnT4MPyP.

Cationic water-soluble porphyrins possessing alkylated pyridine moieties have been shown to bind to DNA. Early work by Fiel, Pasternack and their co-workers involving the porphyrin *meso*-tetrakis-(N-methyl-4-pyridiniumyl) porphine, H$_2$T4MPyP (H2-1), and certain of its metal complexes, revealed that the nature of the porphyrin-DNA interaction was sensitive to metalloporphyrin structure (1, 2). While the free base, H$_2$1, and its Cu^{+2} and Ni^{+2} complexes intercalate between the base pairs of DNA, those porphyrins possessing bulky substituents on the porphyrin framework, or metal ions requiring axial ligands, bind in an outside fashion to DNA. In addition to confirming these initial observations, subsequent investigations by a number of researchers have characterized the DNA binding properties of a variety of different metalloporphyrins (3–8).

0097–6156/89/0402–0074$06.00/0

In this chapter, we examine the mode of binding of MnT4MPyP and certain related complexes to DNA. Since the compound can be activated to produce DNA strand scission, the ability of MnT4MPyP to act as a probe in footprinting experiments involving the antiviral agent netropsin is also presented and discussed.

Experimental

The metalloporphyrins used in the study were prepared as previously described (*9, 10*). Isolation, end-labeling, and sequence analysis of the 139-base pair *Hind*III/*Nci*I restriction fragment of pBR322 DNA was described elsewhere (*9, 11*). The cleavage reactions involving the 3' (position 33) and 5' (position 30) end-labeled fragments and the porphyrins were described earlier (*9,10*). Conditions for cleavage of the 5' end-labeled fragment by MnT4MPyP were identical to those earlier described except that the activating agent was KO_2. Quantification of autoradiographic data was by linear scanning microdensitometry (*12*).

Absorption Studies. All optical studies were carried out in the buffer 50 mM Tris•Cl, 0.1 mM EDTA (pH 7.5). Porphyrin concentration was determined using ε at 460 nm of 9.2×10^4 $M^{-1}cm^{-1}$. Concentrations, in base pairs, for the polymeric DNA's (Boehringer Mannheim) were determined using the base pair molar extinction values of 1.68×10^4 at 254 nm for poly dAdT•poly dAdT (*13*); 1.31×10^4 at 260 nm for poly dGdC•poly dGdC (*14*); 1.38×10^4 at 251 nm for poly dIdC•poly dIdC (*15*). The molar extinction coefficient for the oligonucleotide (CCCTAA)$_3$ of 7.4×10^3 (phosphate) was calculated using published methods (*16*). The oligomer was a gift from M. Schechtman.

Temperature Dependence of MnT4MPyP-Cleavage of a 139-Base Pair Restriction Fragment. For each cleavage reaction, the DNA was placed at the desired temperature (5, 20, 40, 50, 60, 70, 80, or 90 °C) for 10 min before addition of MnT4MPyP, followed by additional equilibration for 15 min. Addition of 2 μl of a 1 mM solution of Oxone (KHSO$_5$) initiated DNA strand scission. Each reaction was allowed to proceed for 7 min, at which time it was terminated using an 8 M urea solution containing 0.025% bromphenol blue and xylene cyanol. All reactions were carried out in a buffer consisting of 50 mM Tris•Cl, 8 mM MgCl$_2$, and 2 mM CaCl$_2$ (pH 7.5), in a total volume of 8 μl. The final concentrations of the various components in the system were: MnT4MPyP, 1 μM; DNA (calf thymus), 193 μM; DNA (labeled fragment), ~1 μM; KHSO$_5$, 250 μM. The reaction products were loaded onto a 12% denaturing gel, and electrophoresed for 2 hr at 150 W. Visualization of the DNA was by autoradiography at –20 °C using Kodak X-Omat AR film without an intensifying screen.

Porphyrin-Mediated DNA Strand Scission

The versatile catalytic nature of the metalloporphyrins suggested that certain metal complexes of $H_2 1$ might be able to facilitate DNA strand scission. This was first investigated by Fiel and co-workers (17) who showed that $Fe^{3+}1$ in the presence of thiols produced DNA breakage. Subsequent investigations by Ward et al. (18, 19) revealed that M-1, where M is Fe^{3+}, Mn^{3+}, or Co^{3+}, in the presence of ascorbate, superoxide ion, iodosobenzene, or $KHSO_5$ could produce DNA breakage. Comparisons between the rates of porphyrin autodestruction and the rates of strand scission of covalently closed circular PM2 DNA further indicated that the metalloporphyrin remains intact during the cleavage process and that it is the species responsible for strand scission. Analysis of the porphyrin cleavage sites, on a 139-base pair restriction fragment of pBR322 DNA using sequencing methodology, revealed that the porphyrin produces a single major cleavage site within its binding site. Analogy with cytochrome P-450 activation chemistry suggests (20) that unlike MPE•Fe(II) and Fe(II) EDTA, which are believed to cleave DNA via a diffusible radical mechanism (21, 22), the cationic metalloporphyrins break DNA through a high-valent oxo intermediate. When produced on or in the vicinity of DNA, this species attacks the deoxyribose moiety, ultimately producing a break in the sugar phosphate backbone of the polymer. While the exact nature of the chemistry at the strand scission site may depend on metalloporphyrin structure and method of activation, MnT4MPyP in the presence of $KHSO_5$ appears to yield a single product associated with the 5' terminus adjacent to the site of scission. However, electrophoretic analysis shows that the product migrates at a slower rate than its DNase I-produced counterpart, indicating that strand scission does not leave a phosphate group on the 5' end of the DNA strand adjacent to the site of scission. Related experiments involving a 5' end-labeled segment of pBR322 DNA, MnT4MPyP, and KO_2 indicate that the group on the 3' end adjacent to the site of scission is likely a phosphate moiety (18).

Structure of the Porphyrin-DNA Complex

In the absence of x-ray crystallographic data on a porphyrin-DNA complex, our understanding of the nature of the binding process rests on the results of a variety of physiochemical studies. The porphyrin macrocyclic framework possesses four cationic charges and is devoid of hydrogen bond donor or acceptor groups which could be used in binding to DNA. Studies with outside-binding porphyrins related to MnT4MPyP show that the interaction is largely electrostatic in nature, and, as expected, is sensitive to ionic strength (23). The exact nature of the ligand(s) occupying the axial site(s) of MnT4MPyP under the conditions of most DNA binding studies are unknown. Cl^-, OH^-, or H_2O are likely candidates. In addition to blocking intercalation, at least two of these groups, OH⁻ and H_2O, may serve as DNA donor/acceptor hydrogen bond sites with DNA.

Although metalloporphyrin binding sites on DNA have been found using DNase I footprinting analysis (7, 24), these can most easily be identified by activating the porphyrin to cause DNA strand scission. As is evident from Figure 1, all sites of type (A•T)₃, at positions 66, 86, 95, 99, and 147 of a 139-base pair restriction fragment of pBR322 DNA, are porphyrin binding sites. For contiguous runs of adenine and thymine greater than three, e.g. at positions 46–50, 56–62, 89–92, and 156–159, the porphyrin cleavage pattern can be explained by porphyrin binding to and cleaving at each trinucleotide site, which overlaps with one or more adjacent trimers. Replacing the four methyl groups on the pyridine moieties of the macrocycle with bulkier groups, as in Mn-2, does not lead to a significant change in the DNA cleavage pattern of the porphyrin (Figure 1). However, if steric bulk is introduced above and below the plane of the porphyrin ring, as with the atropisomer αααβ (9), the DNA cleavage pattern is altered. The manganese complex of this atropisomeric porphyrin cleaves at AT as well as GC sites of DNA. Thus, while substituents directed in the porphyrin plane do not influence DNA sequence recognition, steric hindrance above and below the plane significantly alters specificity.

The site of porphyrin binding is the minor groove of DNA. As is shown in Figure 1, the cleavage events on opposing DNA strands for MnT4MPyP are offset ~2 base pairs in the 3' direction of the 139-base pair restriction fragment. This effect has been observed for other cleaving agents which recognize the minor groove of DNA (22). The fact that the DNA binding sites of the porphyrins can be easily detected by DNase I is also a strong indication that binding takes place in the minor groove. X-ray analysis has shown that this enzyme attacks DNA from its minor groove (25).

Linear dichroism studies involving ZnT4MPyP have shown that the porphyrin is oriented ~67° relative to the helix axis and that binding may be disturbing the Watson and Crick base pairing at the interaction site (6). In light of the high AT specificity and the fact that the binding/cleaving pattern is insensitive to the nature of the substituents on the porphyrin periphery (Figure 1), the porphyrin is likely binding "end-on" in a melted or partially melted region of DNA. Since an AT base pair possesses only two hydrogen bonds and melts at a lower temperature than does a GC base pair, which possesses three hydrogen bonds, the AT specificity is probably related to the ease of DNA melting at the interaction site. This model is also supported by optical studies involving MnT4MPyP. As is shown in Figure 2, the porphyrin Soret band experiences a modest decrease in intensity in the presence of either poly dGdC•poly dGdC or single-stranded DNA. While it is difficult to correlate these optical changes with a precise binding model, the optical effects for poly

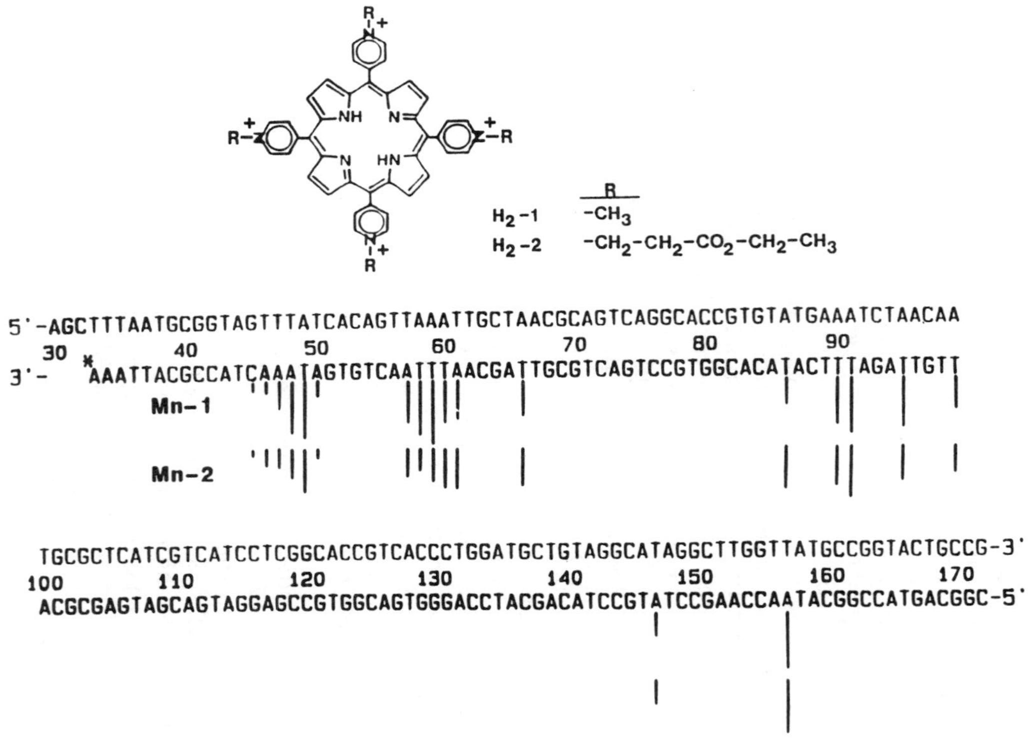

Figure 1. The cleavage sites in histogram format for two cationic manganese porphyrin complexes on a 139-base pair restriction fragment of pBR322 DNA are shown. Due to low resolution of the sequencing gel, the cleavage sites at positions 147 and 157 are estimated.

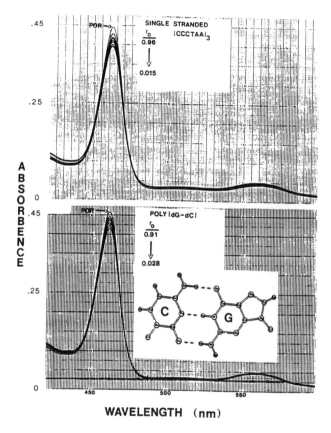

Figure 2. The optical changes which take place in the absorption spectrum of MnT4MPyP in the presence of the single-stranded DNA oligomer d(CCCTAA)3 (top) and double-stranded poly dGdC•poly dGdC (bottom) are shown. In addition to the porphyrin spectrum in the absence of DNA, various spectra at different porphyrin-to-DNA ratios (in base pairs), are shown.

dAdT•poly dAdT and poly dIdC•poly dIdC, which are double stranded but possess only two hydrogen bonds between bases, are dramatically different from poly dGdC•poly dGdC and single stranded DNA. As is shown in Figure 3, the Soret band of MnT4MPyP increases in intensity in the presence of these two polymers, reinforcing the model that melting or partial melting is important in the binding process.

In order to determine if steric effects due to the exocyclic 2-amino group of guanine are also important in the mechanism, the DNA cleavage pattern of MnT4MPyP as a function of temperature was examined. At temperature below 80 °C, cleavage of the 139-base pair restriction fragment by MnT4MPyP/KHSO$_5$ occurs, as expected, only at AT sites (Figure 4). However, above 80 °C, cleavage in GC- as well as AT-rich regions of the polymer is observed (Figure 4). The simplest explanation for cleavage at GC sites at high temperatures is that the thermal energy melts or partially melts DNA in these regions, making them susceptible to porphyrin binding and cleavage. This observation strongly suggests that the steric effect of the exocyclic amino group of guanine in the minor groove is not an important factor in blocking binding at GC sites, but rather that ease of DNA melting determines which sites will be attacked by the compound.

MnT4MPyP as a Footprinting Probe

Footprinting analysis is a widely-used technique for identifying the binding sites of drugs and proteins on DNA molecules derived from natural sources. The method requires an end-labeled segment of DNA which is pre-equilibrated with a DNA binding ligand prior to exposure to a cleavage agent. If the cleavage agent is DNase I or the synthetic compound MPE•Fe(II) (22), which cleave at many sites on DNA, the ligand binding locations can be identified by analyzing the digest products using a DNA sequencing gel. Since cleavage is reduced at ligand binding sites, the locations of binding appear as omissions among the analyzed oligonucleotide products.

In addition to simply mapping the locations of drug and protein binding sites on DNA, footprinting analysis can be used to measure binding constants as a function of sequence. In quantitative footprinting studies involving proteins and the antiviral agent netropsin, DNase I has been shown to yield valid binding information (26–28). Although MnT4MPyP exhibits high specificity for AT sites and thus is not a true footprinting probe, it can be used to monitor site loading events for drugs such as netropsin and distamycin which exhibit high affinities for AT rich regions of DNA.

Key to obtaining binding constants from footprinting data is translating autoradiographic spot intensities into a valid binding isotherm for a given site. This in turn requires a clear understanding of the various events taking place in the experiment, and development of theory useful

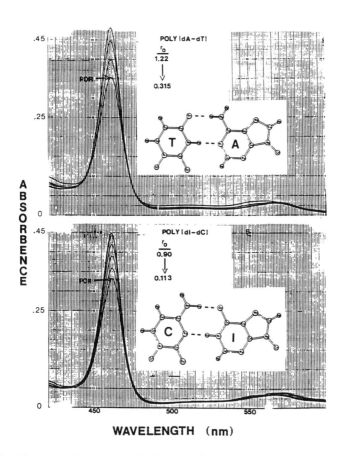

Figure 3. The optical changes which take place in the absorption spectrum of MnT4MPyP in the presence of double-stranded poly dAdT•poly dAdT (top) and poly dGdC•poly dGdC (bottom). See the caption for Figure 2 for additional information.

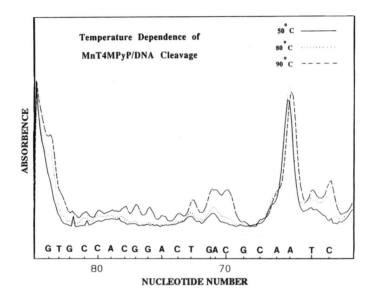

Figure 4. Densitometric scans of autoradiograms obtained from MnT4MPyP/KHSO5 cleavage of a 139-base pair fragment of pBR322 DNA are shown. The various scans are from reactions at 50 °C (———), 80 °C (•••), and 90 °C (- - - -). The sequence and numbering system are given in Figure 1. The radiolabel is at position 33(A) of the fragment.

for quantitative analysis. Although the details of quantitative foot-printing studies with MnT4MPyP and netropsin can be found elsewhere (*19,29*), two important points concerning the ability of this cleavage agent to report quantitative information in the footprinting experiment will be made here.

Site Exclusions. The relationship between the netropsin binding sites and porphyrin binding/cleavage sites on a 139-base pair fragment of pBR322 DNA is shown in Figure 5. In analyzing the netropsin/ MnT4MPyP system, we considered how a drug molecule, which is four nucleotides long, can exclude a porphyrin from its trinucleotide binding site. For the isolated drug sites of type $(A \bullet T)_4$ at positions 89–92 and 156–159 (Figure 5), occupancy by a netropsin molecule excludes both possible modes of porphyrin binding to the tetranucleotide sequence. If a netropsin site is a contiguous set of base pairs indexed by i, i+1, i+2 and i+3, where i is the nucleotide location closest to the label, porphyrin binding at both i, i+1, and i+2, and i+1, i+2 and i+3, is excluded by drug. In this case, the drug binding constant for binding to i, i+1, i+2 and i+3, denoted by K_i, obeys

$$K_i = c_i/D_0c_f \tag{1}$$

in which c_i is the concentration of sites i at which drug is bound, c_f is the concentration of free sites i, and D_0 is the concentration of free drug. The fraction of site i which is covered by drug, f_i, is then given by

$$f_i = K_iD_0/(1+K_iD_0) \tag{2}$$

Since binding of drug to an isolated tetramer blocks both modes of porphyrin binding to the tetramer, the fraction of the porphyrin binding site starting at i which is available to be occupied by porphyrin, v_i, is the same as for the site starting at i+1, i.e.

$$v_i = v_{i+1} = 1 - \frac{K_iD_0}{1 + K_iD_0} \tag{3}$$

The situation with the $A \bullet T$ pentamer at positions 46–50 (Figure 5) is similar except that drug may occupy two different positions within the sequence, either of which excludes the binding of the porphyrin to all three of its binding sites within the segment. The probability that drug covers the sites from i through i+3 is $K_iD_0(1 + K_iD_0 + K_{i+1}D_0)^{-1}$, and the probability that drug covers the sites from i+1 through i+4 is $K_{i+1}D_0(1 + K_iD_0 + K_{i+1}D_0)^{-1}$. Since either situation will prevent the binding of porphyrin,

$$v_i = v_{i+1} = v_{i+2} = 1 - K_iD_0 (1+K_iD_0 + K_{i+1}D_0)^{-1} - K_{i+1}D_0 (1 + K_iD_0 + K_{i+1}D_0)^{-1} = 1/(1 + K_iD_0 + K_{i+1}D_0) \tag{4}$$

For the A-T heptamer located at positions 56–62 (Figure 5) the situation is more complicated. The size of the netropsin binding site on DNA is ~4 base pairs, and the size of the porphyrin binding site is ~3 base pairs

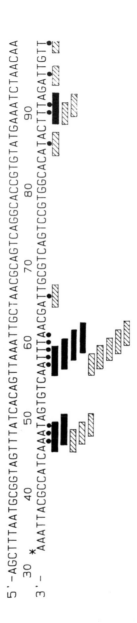

Figure 5. The sites of netropsin binding (filled boxes), porphyrin binding (crosshatched boxes) and porphyrin cleavage (•) on the sequence of the 139-base pair restriction fragment are shown. (Reprinted from ref. 29. Copyright 1989 American Chemical Society.)

(*18, 30*), so drug binding to i+1 through i+4, or i+2 through i+5, will exclude porphyrin binding to any of its five possible binding sites within the sequence. Porphyrin binding to its three interior sites is excluded by drug bound anywhere in the heptamer so the fraction of these sites at which porphyrin could be bound is

$$v_{i+1} = v_{i+2} = v_{i+3} = (1 + D_0 \sum_{j=0}^{3} K_{i+j})^{-1} \qquad (5)$$

which is the probability that no drug is bound to the heptamer. However, it is theoretically possible to have the porphyrin and the drug simultaneously occupy the heptamer if the drug is bound to i through i+3 while the porphyrin occupies i+4 through i+6. An equivalent situation exists for porphyrin occupation of the lowest numbered sites within the heptamer (Figure 5). Although electrostatic effects would discourage simultaneous occupations on the same DNA restriction fragment (both the porphyrin and the drug are cations), we did not introduce any correction for anticooperativity in the quantitative analysis of the netropsin/MnT4MPyP footprinting data, because the simultaneous binding situation is so unimportant among the binding events.

$$v_i = 1 - D_0 \sum_{j=0}^{2} K_{i+j}(1 + D_0 \sum_{j=0}^{3} K_{i+j})^{-1} = D_0 K_{i+3} (1 + D_0 \sum_{j=0}^{3} K_{i+j})^{-1} \qquad (6)$$

Similarly,

$$v_{i+4} = D_0 K_i (1 + D_0 \sum_{j=0}^{3} K_{i+j})^{-1} \qquad (7)$$

There are no A-T sequences of length greater than seven on the *Hind*III/ *Nci*I restriction fragment and the complications just discussed do not occur for A-T segments of smaller length.

Reporting of Site Occupancy By the Cleavage Agent

In the quantitative footprinting experiment, the change in the amount of cleavage at a particular site is used to measure the fraction of that site which is occupied by a drug molecule, f_i. In studies involving DNase I, it was assumed that the cleavage rate at site i is proportional to v_i, calculated by equations like (5–8). This assumption yielded reasonable binding constants for netropsin, and the lac and trp repressors toward their interaction sequences (*26–28*). Since the enzyme binds and cleaves at nearly every nucleotide position, it "sees" a series of contiguous binding sites. If the barrier height between sites is less than kT, the enzyme, similar to restriction endonucleases (*31*), can easily transit in a one-dimensional diffusional process between sites on the DNA lattice.

Effectively, it is not bound to a specific site, but rather to the entire fragment.

If the probe specificity is high, as is the case with MnT4MPyP, the barrier height for transiting from a specific site, e.g. an isolated trimer of type $(A \bullet T)_3$, to an adjacent site which does not bind porphyrin, is likely to be high. In this case there is a classic site-specific equilibrium between bound and unbound cleavage agent, and the cleavage at a particular site will then not be proportional to $1-v_i$. This can be shown using a simple example.

Let K_i and Q_i represent the binding constants for drug and probe at an isolated site. Let c be the total concentration of such sites, P_t the total probe concentration, and D_b and P_b the concentrations of these sites at which drug and probe, respectively, are bound. The cutting rate is proportional to P_b and $v_i = 1 - D_b/c$. The equilibrium expression

$$Q_i = \frac{P_b}{(P_t - P_b)(c - D_b - P_b)} \tag{8}$$

is solved to give

$$P_b = \frac{1}{2}c(v_i) + \frac{1}{2}P_t + (2Q_i)^{-1} - \{c(v_i) + [\frac{1}{2}P_t + \frac{1}{2Q_i}]^2 - cP_tv_i\}^{\frac{1}{2}} \tag{9}$$

In general P_b is not proportional to v_i, but if P_t is large compared to c and Q_i^{-1}

$$P_t = c(v_i) - \frac{c(v_i)}{P_tQ_i} - \frac{[(Q_i)^{-1} - c(v_i)]^3}{4P_t^2} + \ldots \tag{10}$$

This means that, for the footprint to report the fraction of binding site unoccupied by drug, the probe binding constant and the amount of probe must be large enough to saturate the site with probe. For the quantitative experiments involving MnT4MPyP and netropsin, c and P_t are ~50 µM and 1.3 µM respectively (*19, 29*). Based on studies involving related porphyrins, the binding constant toward calf thymus DNA can be estimated to be ~10^7 M^{-1} (*6, 23*). Thus, for these experiments P_b is not proportional to v_i. This point must be addressed in developing the theory for quantitative experiments involving MnT4MPyP. In the model developed for analyzing the footprinting data, it is not necessary to consider possible drug-drug and porphyrin-porphyrin interactions which may be present in the system. Since the porphyrin and drug concentrations are relatively low in the experiment, the probability that a single DNA restriction fragment possesses more than one of either is also low. Thus, possible homo-cooperative effects, porphyrin-porphyrin or drug-drug interactions, are remote. An additional factor in this case is that the drug binds without greatly distorting DNA (*30*). This property would minimize the opportunity to influence the binding to other drugs *via* propagation of allosteric effects through the DNA helix.

Possible cooperative effects between drug and the porphyrin were also not considered. In view of the sizes of the drug and porphyrin, nearest-neighbor cooperative effects can only take place for the heptamer located at positions 56–62 and then only for two porphyrin cutting sites within the heptamer. If cooperativity were to take place at these positions, its presence would have only a modest impact on the amounts of fragments produced in the cleavage process. Since netropsin increases the melting point of DNA, it may be possible to alter porphyrin binding at other non-overlapping strong netropsin sites on the 139-mer. However, the strong sites are well separated on the polymer, and changes in DNA melting due to drug binding would not be expected to be propagated over long distances through the double helix. Evidence that this is true can be obtained by observing the porphyrin cleavage at all isolated trimers of type $(A \bullet T)_3$ which can bind porphyrin but cannot bind drug (*19, 29*). The cleavage rates at all of these sites are the same as drug is added to the system, indicating that binding events can be treated in an independent non-cooperative manner.

As earlier outlined by us (*32*), the reporting of site loading information in the footprinting experiment is independent of the detailed cleavage mechanism of the probe and the binding kinetics of the ligand. If the system is at equilibrium before cleavage and remains so during the time course of the reaction, equilibria are being measured (*28*). In thinking about the various events taking place in the footprinting experiment, it may be useful to visualize a row of houses on a city street. The houses, which correspond to the nucleotide positions of DNA, have closed doors, which are randomly checked for their locked and unlocked status by a visitor who roams the street (the footprinting probe). If there is no drug present in the experiment, all of the doors of the houses will be unlocked and the visitor is free to enter, and ultimately exit, leaving the door open. When drug is added to the system, certain doors will be locked thus preventing the visitor from entering those houses. An observer (the experimenter), wishing to know which houses were unlocked, is not concerned about whether the visitor opened the door with the left or right hand, or with either foot (the cleavage mechanism), but simply notes which houses have their doors remaining open (the cut site). DNase I can check any house, but the porphyrin can only check, for example, red ones. Clearly, no information is available on houses which are not red if the visitor acts like MnT4MPyP; those doors will certainly be closed.

Conclusions. In this chapter we have summarized what is known concerning the binding/cleavage mechanism of the DNA-binding metalloporphyrin MnT4MPyP. The bulk of the evidence which has been accumulated to date suggests that the compound binds in the minor groove in a melted or partially melted region of DNA. Preliminary analysis of quantitative footprinting studies with MnT4MPyP and the antiviral agent netropsin revealed that porphyrin and drug site exclusions must be part of the model leading to the measurement of binding constants. In addition, the site-specific nature of DNA cleavage by MnT4MPyP leads to autoradiographic spot intensities which do not directly reflect the occupancy of a drug binding site on DNA by netropsin. Both factors must be addressed in extracting valid binding constants from footprinting data involving MnT4MPyP.

Literature Cited

1. Fiel R. J.; Carvlin M. J.; Byrnes R. W.; Mark, E. H. in "Molecular Basis of Cancer, Part B: Macromolecular, Recognition, Chemotherapy and Immunology"; Rein, R., Ed., Alan R. Liss: New York, **1985**, p. 215.
2. Pasternack, R. T.; Autebi, A.; Ehrlich, B.; Sidney, D. *J. Mol. Catal.* **1984**, *23*, 235.
3. Kelly, J. M.; Murphy, M. J.; McConnell, D. J.; OhUigin, C. *Nucleic Acids Res.* **1985**, *13*, 167.
4. Strickland, J. A.; Banville, D. L.; Wilson, W. D.; Marzilli, L. G. *Inorg. Chem.* **1987**, *26*, 3398.
5. Bortolini, O.; Ricci, M.; Meunier, B.; Friant, P.; Ascone, L.; Goulou, J. *Nouv. Chim.* **1986**, *10*, 39.
6. Geacintov, N. E.; Ibanez, V.; Rougee, M.; Bensasson, R. V. *Biochemistry* **1987**, *26*, 3087.
7. Ford, K.; Fox, K. R.; Neidle, S.; Waring, M. J. *Nucleic Acids Res.* **1987**, *15*, 2221.
8. Doan, T. L.; Perrouault, S.; Chassignol, M.; Thuong, N. T.; Helene, C. *Nucleic Acids Res.* **1987**, *15*, 8643.
9. Bromley, S. D.; Ward, B. W.; Dabrowiak, J. C. *Nucleic Acids Res.* **1986**, *14*, 9133.
10. Raner, G.; Ward, B. W.; Dabrowiak, J. C. *J. Coord. Chem.* **1988**, *19*, 17.
11. Lown, J. W.; Sondhi, S. M.; Ong, C.-W.; Skorobogaty, A.; Rich, N.; Calvin, P. H. V.; Vournakis, J. N. *Nucleic Acids Res.* **1986**, *14*, 489.
12. Dabrowiak, J. C.; Skorobogaty, A.; Rich, N.; Calvin, P. H. V.; Vournakis, J. N. *Nucleic Acids Res.* **1986**, *14*, 489.
13. Schmechel, D. E. V.; Crothers, D. M. *Biopolymers* **1971**, *10*, 465.
14. Müller, W.; Crothers, D. M. *J. Mol. Biol.* **1968**, *35*, 251.
15. Grant, R. C.; Harwood, S. J.; Wells, R. D. *J. Am. Chem. Soc.* **1968**, *90*, 4474.
16. Handbook of Biochemistry and Molecular Biology 3rd ed. (Fasman, G.D., ed.) Vol. 1, CRC Press, Cleveland, Ohio, **1975**, p. 589.
17. Fiel, R. J.; Beerman, T. A.; Mark, E. H.; Datta-Gupta, N. *Biochem. Biophys. Res. Commun.* **1982**, *107*, 1067.

18. Ward, B. W.; Skorobogaty, A.; Dabrowiak, J. C. *Biochemistry* **1986**, *25*, 6875.
19. Ward, B. W.; Rehfuss, R.; Dabrowiak, J. C. *J. Biomol. Struct. Dynamics* **1987**, *4*, 685.
20. Groves, J. T.; Nemo, T. E. *J. Am. Chem. Soc.* **1983**, *105*, 6243.
21. Tullius, T. D. *Nature* **1988**, *332*, 663.
22. Dervan, P. B. *Science* **1986**, *232*, 464.
23. Carlin, M. J.; Datta-Gupta, N.; Fiel, R. J. *Biochem. Biophys. Res. Commun.* **1982**, *108*, 66.
24. Ward, B. W., Skorobogaty, A.; Dabrowiak, J. C. *Biochemistry* **1986**, *25*, 7827.
25. Suck, D.; Yohm, A.; Oefner, C., *Nature* **1988**, *332*, 464.
26. Brenowitz, M.; Senear, D. F.; Shea, M. A.; Ackers, G. K. *Proc. Natl. Acad. Sci. USA* **1986**, *83*, 8462.
27. Carey, J. *Proc. Natl. Acad. Sci. USA* **1988**, *85*, 975.
28. Ward, B. W.; Rehfuss, R.; Goodisman, J.; Dabrowiak, J. C. *Biochemistry* **1988**, *27*, 1198.
29. Dabrowiak, J. C.; Ward, B.; Goodisman, J. C. *Biochemistry* **1989**, *28*, 3314.
30. Kopka, M. L.; Yoon, D.; Goodsell, P.; Pjura, R. E.; Dickerson, R. E. *Proc. Natl. Acad. Sci. USA* **1985**, *82*, 1376.
31. Winter, R. B.; von Hippel, P. H. *Biochemistry* **1981**, *20*, 6948.
32. Goodisman, J.; Dabrowiak, J. C. *J. Biomol. Struct. Dynamics* **1985**, *2*, 967.

RECEIVED May 11, 1989

Chapter 6

Searching for Metal-Binding Domains

Jeremy M. Berg

Department of Chemistry, Johns Hopkins University, Baltimore, MD 21218

In recent years several families of proteins with nucleic acid binding or gene regulatory activities have been discovered to contain short stretches of amino acids that bind metal ions through invariant cysteine and/or histidine residues. These have been termed metal-binding domains. A wide spectrum of methods for identifying additional members of these families have been developed. Four such methods are: (i) screening DNA libraries by hybridization to single stranded nucleic acid probes that are derived from known metal-binding domains; (ii) examining protein sequence libraries via computer for patterns of cysteines and/or histidines; (iii) developing genetic screens for metal ion-dependent phenotypes; and (iv) investigating the metal ion dependence of biochemical activities such as nucleic acid binding. Examples of each of these approached are described.

ANALYSIS OF THE DEDUCED AMINO ACID SEQUENCE of a gene regulatory protein from *Xenopus laevis* called Transcription Factor IIIA (TFIIIA) led to the proposal that this protein contains nine small structural domains each of which is organized around a bound zinc ion (1,2). This hypothesis was based on the discovery that the sequence contained nine tandem regions that match the consensus sequence (Tyr,Phe)-X-Cys-$X_{2,4}$-Cys-X_3-Phe-X_5-Leu-X_2-His-$X_{3,4}$-His-X_{2-6}. Furthermore, it was shown that the protein contained 7–11 zinc ions when a specific protein-RNA complex was isolated in the absence of chelating agents (1). It was proposed that each short sequence bound a zinc ion through the invariant cysteine and histidine residues. The structural domains produced were termed "zinc fingers" (1). A variety of data now supports this proposal including the discovery of numerous other proteins that also contain regions that closely match the consensus sequence above (3–6). In addition, several other classes of genes have recently been sequenced to yield deduced

0097–6156/89/0402–0090$06.00/0

amino acid sequences that are highly suggestive of metal-binding domains. These discoveries have inspired the development of methods for intentionally searching for evidence of such domains in other proteins.

Searching Based on Nucleic Acid Hybridization

The level of sequence similarity between the "zinc finger" domains suggested that it might be possible to deliberately clone genes encoding such regions based on differential hybridization of nucleic acid fragments that are approximately complementary in sequence. Such fragments might be derived from another "zinc finger"-encoding gene, or they might be oligonucleotides that have sequences designed based on a particular amino acid sequence with allowances for the degeneracies in the genetic code. The chosen DNA fragments are then radiolabeled and used to probe a DNA library. This is accomplished by transferring the DNA from bacterial colonies or bacteriophage plaques onto filters, denaturing the bound DNA to expose the potential hydrogen bonding sites, hybridizing the probe onto the filters, and washing the filters under appropriate conditions of salt concentration and temperature so that the probe remains bound only to those sites for which it has the highest affinity (and, therefore, the highest sequence similarity).

This method has been quite useful for identifying additional members of the "zinc finger" protein family. Initial studies using the "zinc finger" coding region from the *Drosophila Krüppel* gene lead to the isolation of complete or partial open reading frames from *Drosophila* (7), mouse (8), and *Xenopus* libraries (9). One of the most spectacular results from these studies was the isolation of a cDNA clone that appears to encode a protein that contains 37 "zinc finger" sequences from a *Xenopus* library (9)! Such studies have also provided evidence for the presence of other "zinc finger" protein genes in a variety of other DNA libraries. It appears that all eukaryotic genomes contain multiple "zinc finger" en-coding regions, whereas a representative prokaryote, *E. coli*, does not (7,8).

A more recent study utilized a synthetic oligonucleotide probe designed from the relatively conserved sequence found in the linker between "zinc finger" domains in certain proteins (10). This probe was used to screen *Xenopus* libraries to yield 14 different clones encoding a total of 109 "zinc finger" sequences. These proteins share the conserved linker region used in the screening but also contain other conserved features in the "zinc finger" regions themselves.

A disadvantage of this approach is that while numerous open reading frames are identified, there is little information concerning the functions of the proteins that are encoded by these genes. One experimental approach that has yielded interesting information is the study of messenger RNA levels at different developmental stages. With the family of *Xenopus* genes, it appears that individual members of the family are actively expressed at different times during early embryogenesis (10). It will be tremendously exciting when more details of these gene expression networks are in hand.

Searching Computer-Based Protein Sequence Libraries

The amount of protein sequence data available has increased rapidly in recent years due to the development of gene cloning and DNA sequencing methods. Much of these data are compiled in the form of computer-based sequence data bases. Such data bases may be searched for the presence of particular patterns of amino acids.

The most straightforward approach for searching a computer-based sequence library involves the use of a given amino acid sequence as a probe. For example, the sequence of TFIIIA might be compared with all of the other sequences in a library using an alignment procedure to arrange the probe sequence against each of the members of the library, and a scoring procedure to assess the degree of similarity. Such methods work quite well for proteins that are relatively close in sequence, assuming that suitable procedures are employed for allowing gaps to occur in the alignment. The procedure must be flexible enough to account for small deletions and insertions that may be present in one sequence but not the other, but not so loose that unrelated sequences can be aligned artificially well. These methods have been employed to discover that a number of recently cloned and sequenced genes encode proteins that contain sequences that match the "zinc finger" consensus noted above.

An alternative strategy involves the use of a more general template for searching a computer-based library. Such a template explicitly includes those residues that are believed to be functionally or structurally important, and often allows more significant gapping than is possible using a single probe sequence. A search for potential metal-binding domains has been performed using the templates $Cys-X_{2-4}-Cys-X_{2-15}-a-X_{2-4}-b$ and $a-X_{2-4}-b-X_{2-15}-Cys-X_{2-4}-Cys$, where X may be any amino acid and a and b may be either Cys or His (11). This template was chosen based on analogy with the zinc-binding domains from TFIIIA and on an analysis of short metal-binding regions from crystallographically characterized metalloproteins. This search identified five classes of proteins that had been implicated in nucleic acid binding or gene regulation. Included among these are the retroviral core nucleic acid binding proteins. These proteins contain either one or two sequences of the form $Cys-X_2-Cys-X_4-His-X_4-Cys$. The potential metal-binding residues are the only ones that are completely conserved in these regions. Furthermore, these sequences have other features, such as relatively well conserved Gly residues, that are suggestive with regard to their role as metal-binding sites (11). Recently reported data obtained using a new zinc "blotting" technique provides the first evidence that these proteins do, in fact, bind zinc ions under certain conditions (12). Further studies are clearly necessary to confirm the role of these sequences in metal binding and to correlate the presence of metal ions in the proteins with functional properties.

The hybridization and the computer search methods are linked by a common feature: the need to control stringency. In either case, it is quite easy to choose parameters inappropriately so that either no new sequences are detected or a large number of "false positives" are found. It is often necessary to try a range of parameters in order to optimize the new information obtained by the search. Further experience with both types of procedure should make this process more systematic.

Genetic Approaches to Identifying Metal Binding Domains

The characterization of functionally defective mutants has played a role in elucidating the roles of zinc ion in two systems. In these studies, mutants were selected based on their inability to grow under certain conditions. The mutations that had occurred were mapped using genetic and DNA sequencing techniques. Subsequent studies revealed that the defects could be largely corrected by the addition of appropriate concentrations of zinc to the growth medium. The success of this approach is based on the assumption that the wild type proteins have evolved such that they are functionally saturated with zinc under normal growth conditions, so that no zinc-dependent activity is observed. Growth under metal-depleted conditions leads to a variety of effects that are generally irrelevant to the system under study. However, the generation of specific mutants can yield proteins that have lowered effective affinities for metal ions such that they are not saturated under normal growth conditions. Thus, metal-dependent behavior is observed if the medium is supplemented with the appropriate metal ions such that increased levels of metal are bound by the mutated proteins and changes in activity occur.

Dramatic results have been obtained using this approach for the study of the GAL4 protein from the yeast *Saccharomyces cerevisiae* (*13,14*). This protein regulates the transcription of a variety of genes required for the utilization of galactose. Mutants were selected based on their inability to grow on a galactose-based medium. The majority of the mutations were found to occur in a short region at the amino terminus of the protein (*13*). This region had already been implicated as the DNA-binding domain (*15*) and contains a sequence of the form $Cys-X_2-Cys-X_6-Cys-X_6-Cys-X_2-Cys-X_6$-Cys that is common to a class of regulatory proteins from fungi (*16*). It had been suggested that this region corresponds to a metal-binding domain (*13,17*). Strong support for this proposal was obtained from further studies of the mutants. It was discovered that the ability of one of the mutants to grow on galactose was restored by addition of zinc (*14*). In particular, it was found that placement of a crystal of $ZnCl_2$ on the edge of a Petri dish containing galactose medium and plated with the mutant strain led to the appearance of a halo of yeast growing at a certain distance around the crystal. This effect was unique to zinc, and was observed only for one class of mutant involving modification of a proline residue that is conserved in the six known proteins in this class.

These observations may be interpreted as follows. The region is, in fact, a zinc-binding domain that is involved in DNA binding. Modification of the conserved proline residue yields protein that has a significantly reduced affinity for zinc such that a substantial amount of the zinc-containing protein is formed only in the presence of higher concentrations of zinc than are normally present. The other mutations presumably act either by affecting some property required for biological activity, such as a residue involved in contacting DNA, or by decreasing the affinity for zinc to a level that it cannot be restored by supplementation of the medium.

A second protein for which genetic studies has aided in characterization of the role of zinc is the helix-destabilizing protein from bacteriophage T4, often called the gene 32 protein (*18*). This protein contains a sequence of the form $Cys-X_3-His-X_5-Cys-X_2-Cys$ that was proposed as a metal-binding site based on the computer search described above (*11*). Almost simultaneously, it was discovered that the protein contains one equivalent of zinc as isolated (*19*). These observations led to the reexamination of a number of temperature sensitive mutants involving changes in gene 32 (*18*). It was found that many of the altered properties of the mutants more closely matched wild-type behavior when the growth medium was supplemented with zinc. Furthermore, it was found that many of the mutations occurred in or adjacent to the proposed metal-binding domain. These observations are consistent with the importance of the metal-binding domain for the folding and stability of this protein.

Metal-Dependent DNA-Binding Activity as an Indicator of the Presence of Metal-Binding Domains

Many of the proteins discussed above are sequence-specific DNA binding proteins. For example, TFIIIA specifically recognizes a region of approximately 50 base pairs within 5S RNA genes (*20*). Studies have shown that this specific DNA binding activity requires the presence of zinc within the protein (*21*). Thus, treatment of TFIIIA with chelating agents such as EDTA results in loss of detectable specific binding activity. Furthermore, the binding activity can be restored by addition of zinc to EDTA-treated samples. Similar observations have been made with the human transcription factor Sp1 (*22*). These results suggest that metal dependent DNA binding activity may be characteristic of gene regulatory proteins that contain metal-binding domains.

One experiment that has been particularly useful for demonstrating such effects is the so-called gel shift assay. This technique is based on the differences in mobility on a native electrophoresis gel between a free fragment of DNA and its complex with a protein. The experiment is generally performed as follows. A DNA fragment containing the binding site of interest is prepared and radiolabeled. A portion of this preparation is treated with a potential DNA-binding protein, either in pure form or in a crude extract, and the samples are run on an electrophoresis gel under non-denaturing conditions. The gel is then subjected to autoradiography. If the DNA-binding protein does indeed bind to the DNA fragment used then the radioactivity will be shifted from the position characteristic of the free DNA to a position corresponding to the complex. Advantages of this assay are that it is simple and that it can be used with crude extracts or only partially purified proteins.

This assay has been used to investigate the metal dependence of the DNA binding activities of two proteins called H4TF-1 and H4TF-2 that are involved in transcriptional control of human histone H4 genes (23). These proteins have been partially purified based on their interactions with particular sequences upstream of the histone genes. The proteins show gel shift activity with appropriate DNA fragments. This activity was abolished by treatment of the protein with EDTA or 1,10-phenanthroline. For H4TF-1, the activity could be restored by treatment with zinc but no other metal ion. However, for H4TF-2 both zinc and iron(II) restored gel shift activity. In addition, it was shown that preformed protein-DNA complexes are relatively resistant to the chelating agents. It will be very interesting to see how these activities and the apparent differences in metal specificity can be explained when these proteins are purified and more fully characterized. It should be noted that not all metal-binding domain-containing proteins are sensitive to chelating agents such as EDTA under reasonably mild conditions. Thus, the absence of an effect of EDTA on an activity should not be interpreted as conclusive evidence against the presence of metal ions in the protein.

Conclusions

The above examples illustrate four quite different methods that have been developed in recent years for discovering and characterizing metal-binding domains in proteins involved in nucleic acid binding and gene regulation. It is quite clear from the rapid development of this field over the past few years that such domains play central roles in many gene regulatory systems. Recent experiments have revealed that not all metal-binding domains in proteins involved in nucleic acid binding or gene regulation are directly involved in interactions with nucleic acids; it appears that such domains can simply stabilize the three-dimensional structure of a protein or mediate protein-protein interactions (24). Further investigations of many aspects of proteins that contain metal-binding domains, including their structural and metal-binding properties as well as the mechanisms by which they control gene transcription, are expected to yield many exciting results in the near future.

Note added in proof
The metal-binding activity of the retroviral Cys-X$_2$-Cys-X$_4$-His-X$_4$-Cys region has been directly demonstrated using appropriate peptides (South, T. L. et al. *J. Am. Chem. Soc.* **1989**, *111*, 395, and Green, L. M.; Berg, J. M. *Proc. Natl. Acad. Sci. USA* **1989** (in press).

Literature Cited

1. Miller, J.; McLachlan, A. D.; Klug, A. *EMBO J.* **1985**, *4*, 1609.
2. Brown, R. S.; Sander, C.; Argos, P. *FEBS Lett.* **1985**, *186*, 271.
3. Berg, J. M. *Nature* **1986**, *319*, 264.
4. Vincent, A. *Nucleic Acids Res.* **1986**, *14*, 4385.
5. Klug, A.; Rhodes, D. *Trends Biochem. Sci.* **1987**, *12*, 464.
6. Evans, R. M.; Hollenberg, S. M. *Cell* **1988**, *52*, 1.
7. Schuh, R.; Aicher, W.; Gaul, U.; Côté, S.; Preiss, A.; Maier, D.; Seifert, E.; Nauber, U.; Schröder, C.; Kemler, R.; Jäckle, H. *Cell* **1986**, *47*, 1025.
8. Chowdhury, K.; Deutsch, U.; Gruss, P. *Cell* **1987** *48*, 771.
9. Ruis i Altaba, A.; Perry-O'Keefe, H.; Melton, D. A. *EMBO J.* **1987**, *6*, 3065.
10. Köster, M.; Pieler, T.; Pöting, A.; Knöchel, W. *EMBO J.* **1988**, *7*, 1735.
11. Berg, J. M. *Science* **1986**, *232*, 485.
12. Schiff, L. A.; Nibert, M. L.; Fields, B. N. *Proc. Natl. Acad. Sci. USA* **1988**, *85*, 4195.
13. Johnston, M.; Dover, J. *Proc. Natl. Acad. Sci. USA* **1987**, *84*, 2401.
14. Johnston, M., *Nature* **1987**, *328*, 353.
15. Keegan, L.; Gill, G.; Ptashne, M. *Science* **1986**, *231*, 699.
16. Beri, R. K.; Whittington, H.; Roberts, C. F.; Hawkins, A. R. *Nucleic Acids Res.* **1987**, *15*, 7991, and references therein.
17. Hartshorne, T. A.; Blumberg, H.; Young, E. T. *Nature* **1986**, *320*, 283.
18. Gauss, P.; Krasa, K. B.; McPheeters, D. S.; Nelson, M. A.; Gold, L. *Proc. Natl. Acad. Sci. USA* **1987**, *84*, 8515.
19. Giedroc, D. P.; Keating, K. M.; Williams, K. R.; Konigsberg, W.; Coleman, J. E. *Proc. Natl. Acad. Sci. USA* **1986**, *83*, 8452.
20. Engelke, D. R.; Ng, S. Y.; Shastry, B. S.; Roeder, R. G. *Cell* **1980**, *19*, 717.
21. Hanas, J. S.; Hazuda, D .J.; Bogenhagen, D. F.; Wu, F. Y.-H.; Wu, C.-W. *J. Biol. Chem.* **1983**, *258*, 14120.
22. Kadonaga, J. T.; Carner, K. R.; Masiarz, F. R.; Tjian, R. *Cell* **1987**, *51*, 1079.
23. Dailey, L.; Roberts, S. B.; Heintz, N. *Mol. Cell. Biol.* **1987**, *7*, 4582.
24. Berg, J. M. *Prog. Inorg. Chem.* **1989**, *37*, *143*, and references therein.

RECEIVED May 11, 1989

Chapter 7

Inorganic Reagents as Probes for the Mechanism of a Metal-Responsive Genetic Switch

Betsy Frantz and Thomas V. O'Halloran

Department of Chemistry and Department of Biochemistry, Molecular Biology and Cell Biology, Northwestern University, Evanston, IL 60201

Inorganic reagents can provide detailed structural information concerning the specific interactions of biopolymers in large assemblies. One such assembly of biopolymers controls the expression of mercuric ion resistance genes in bacteria. The central component of this control mechanism is the MerR metalloregulatory protein, a small DNA-binding protein which binds mercuric ion with high affinity, and upon doing so, activates gene expression. This article reviews insights into the configuration of the biopolymers involved in the metal-responsive switching event, and focuses on the utility of two coordination complexes that were developed as tools for molecular biology.

A common cellular response to stress induced by changes in extra-cellular metal ion concentrations is the alteration of expression of genes which encode heavy metal storage, efflux or uptake proteins (1). The central components in several of these metal-responsive genetic mechanisms are metalloregulatory proteins, heavy metal sensor/receptors that can ultimately switch on or off the expression of metal-responsive genes. In order to understand the structure and function of metal-responsive genetic switching mechanisms in general, we have purified components of a mercuric ion responsive switch which activate the expression of mercuric ion resistance genes in bacteria.

The central component of this switch is the MerR metalloregulatory protein, a small DNA-binding protein which binds mercuric ion with high affinity. The MerR protein acts as a repressor in the absence of mercuric ion, and an activator of RNA transcription in the presence of submicromolar levels of mercuric ion. We have

0097–6156/89/0402–0097$06.00/0

reconstructed the switching activity *in vitro* and explored the interactions of various biopolymers that constitute the switch (2).

Important insights into the configuration of the biopolymers at various intermediate stages in the switching mechanism have been obtained using inorganic reagents developed as tools for molecular biology. This article will focus on the use of an established inorganic reaction, and introduce the application of another inorganic reagent, to elucidate the metal-responsive switching mechanism. These and related reagents are just a few of a rapidly growing series of coordination complexes which have been developed as probes of biopolymer structure and function. Our efforts to elucidate the mechanics of the mercuric ion responsive switching mechanism represent just one of many approaches. A variety of mutagenesis (3), *in vivo* transcription and *in vivo* footprinting (4) studies have been conducted on the MerR protein, and these have been reviewed elsewhere (5, 6).

Metalloregulatory proteins can conceivably alter gene expression either by changing the frequency of DNA transcription into mRNA, or by changing the frequency of mRNA translation into protein. The DNA sequences of several other heavy metal receptors have been determined. A few of these exhibit metal-responsive DNA binding activity *in vitro*, including the Fur protein which mediates iron responsive regulation of genes responsible for iron uptake in bacteria (7), and the ACE1 protein, a copper-responsive regulator of metallothionein genes in yeast (8). All of these proteins alter transcription in a metal-responsive manner. Factors which mediate metal-responsive translation have been implicated in iron-responsive regulation of ferritin and transferrin, however no metalloregulatory proteins have yet been identified (9).

The earliest probes of the interaction of the MerR metalloregulatory protein with DNA utilized the enzyme DNase I to cleave the DNA backbone (10). Protein/DNA complexes were formed and then treated with the cleaving enzyme. Interaction of the protein with the DNA protected regions of the sugar-phosphate backbone from cleavage by DNAase I. The DNA sequence at the protected region is readily obtained from DNA sequencing gels and yields a pattern which is referred to as a "footprint." These straightforward protection experiments, while clearly demonstrating that the MerR protein had a specific DNA binding activity, failed to show any difference in the presence or absence of mercury. Later experiments, which utilized dialyzed Hg-MerR protein with a stoichiometry of one mercuric ion per MerR dimer, also showed no difference from the footprints obtained in the absence of mercury (2). DNase I footprints provide a good definition of the boundaries of the protein binding site on the DNA, however they do not provide high resolution information about protein/DNA interaction. DNase I is a large enzyme and it typically does not discriminate between subtle differences in biopolymer contacts within a protein/DNA complex.

Other footprinting procedures, such as dimethylsulfate (DMS) protection, principally provide information about the accessibility of the N7 residue of guanine in the major groove. DMS protection protocols, analogous to the DNase I footprinting experiments described above, provided additional information about the interaction of MerR protein with DNA, but still showed no difference in the presence of the inducer Hg(II). The simple DMS protection results can be deceptive since methylation of critical guanine bases in the protein/DNA complex can lead to dissociation of the complex. In this case, DMS reaction with the free DNA can occur in parallel to DMS reaction with the protein/DNA complex and the resulting footprint is characteristic of a mixture of species. This phenomenon was discovered by separating free DNA from the protein/DNA complexes after the DMS reaction. The DMS footprint of the isolated protein/DNA complex shows much more extensive protection in comparison to the footprint of the unseparated mixture. Several important sites of protection at the center of the operator have not been observed in DMS footprints of MerR obtained when separation of the complexes was not employed (11).

These results compliment the high resolution data obtained from the application of the Tullius hydroxyl radical footprinting protocol, which utilizes $[Fe(EDTA)]^{2-}$ and H_2O_2 in the Fenton reaction to generate a DNA nicking agent (12). The hydroxyl radicals generated by this reaction cleave the DNA backbone. Unlike DMS and DNase I, the reaction of the small, solvent-like hydroxyl radical is not dependent on the local DNA sequence. This lack of specificity leads to a uniform reactivity of all solvent-accessible sugar moieties in the DNA backbone, providing a great deal of information on the sites where the MerR protein prevents interaction with solvent. Combination of the DMS and hydroxyl radical results allowed us to develop a stereospecific model for the interaction of the MerR protein with the DNA template. These results are summarized in Figure 1, where the cylindrical DNA molecule has been projected into a two-dimensional perspective. The parallel lines delineate the major and minor grooves and the outline encloses regions inaccessible to hydroxyl radicals and/or DMS.

As with the DMS and DNase I experiments, the hydroxyl radical probe did not show an effect of mercuric ion in the footprinting pattern. We therefore looked for evidence for a molecular basis of the mercuric responsive switching event in the differential interaction of RNA polymerase (RNAP) with MerR/DNA and Hg-MerR/DNA complexes. In order to transcribe a gene, RNA polymerase must bind to specific DNA sequences found approximately 10 and 35 base pairs upstream of the start of transcription. This region is collectively

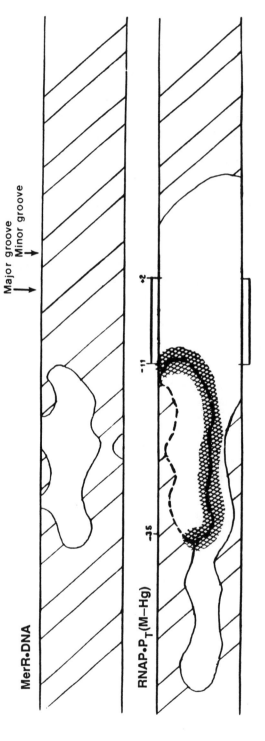

Figure 1. Graphical summary of hydroxyl radical, DMS and KMnO₄ chemical protection experiments on the MerR/DNA (upper) and the RNA polymerase/DNA/MerR-Hg (lower) complexes. The DNA double helix is shown in a two-dimensional projection. Imagine the DNA molecule being cut along the length of the cylinder and flattened out. This projection displays protection sites on the front and back sides of the duplex. The lower complex is a stable and transcriptionally competent intermediate. It is the Hg(II)-induced, 'activated' configuration of the genetic switch. For further details see (2).

referred to as a promoter. Several different stages in the interaction of RNA polymerase with a promoter have been identified by kinetic and footprinting analysis. We are testing the proposal that, in response to submicromolar levels of Hg(II), the MerR protein induces transcription of resistance genes by increasing the rate of RNA polymerase open complex formation at the metal-responsive promoter, P_T.

The terms 'open' and 'closed' refer to discrete complexes of RNA polymerase and DNA molecules containing promoter sequences. A closed complex is initially formed in the interaction of RNA polymerase with the promoter, and the DNA remains double-stranded. A variety of kinetic and physical methods have indicated that the closed complex isomerizes to a stable melted-out intermediate known as the open complex, which is characterized by a region of unpaired DNA bases directly preceding the site of transcription initiation at the +1 site. This region of the promoter spanning the –10 consensus sequence of the bacterial promoter was originally shown by Kierkegaard et al. (13) to be differentially reactive to dimethyl sulfate at cytosine residues which are unpaired in the open complex. Normally, base-paired cytosines would not be modified by DMS under the conditions used for the Maxam and Gilbert guanine-specific reaction. We have found another reagent, potassium permanganate ($KMnO_4$), which readily identifies unpaired thymidine bases in duplex DNA. $KMnO_4$ has been utilized to probe altered DNA structures (14), however this is the first application of $KMnO_4$ to the delineation of single-and double-stranded regions of duplex DNA in a protein/DNA complex (2).

The differential modification by $KMnO_4$ of unpaired thymidines in RNAP open complexes is very rapid in comparison to base-paired regions, resulting in a very low background. We attribute the slowness of the reaction of $KMnO_4$ with base-paired thymidines to steric hindrance arising from adjacent bases in the double strand. Permanganate modification of thymidine has been postulated to occur through oxidation of the C5-C6 double bond to a *cis*-diol (15). This reaction may occur through a manganate ester or metallacycle intermediate, both of which would yield a *cis*-diol product. As shown in Figure 2, initial interaction of the oxoanion above or below the pyrimidine ring would be sterically hindered by an adjacent base in a rigid base-paired duplex. Loss of the Watson-Crick base pairs increases flexibility and accessibility of the heterocyclic bases, and thus single-stranded regions could react more rapidly. We have established reaction conditions compatible with protein/DNA complex formation. Because the –10 region contains a high percentage of A–T nucleotide pairs, modification by permanganate provides a convenient assay for open complex formation.

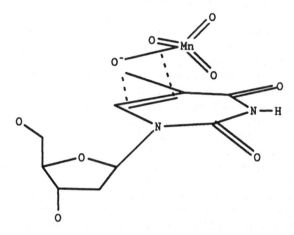

Figure 2. The initial reaction of KMnO₄ with the thymidine can occur through suprafacial attack of the oxyanion at the C5-C6 double bond. If the adjacent base pairs are held parallel in a B-form double helix, they can sterically occlude the C5-C6 double bond and decrease the rate of oxidation to the cis-diol.

The permanganate assay indicated that RNAP forms an open complex with the promoter for the *merR* gene (P_R) *in vitro*, in the absence of the MerR protein. Addition of MerR in the absence of Hg(II) decreased the modification of thymidines in the -10 region suggesting little or no open complex formation at P_R. This is consistent with *in vivo* analysis of promoter fusions which indicate that MerR represses this promoter (*16*). Addition of Hg(II) and MerR greatly enhanced the modification of all of the thymidines between -12 and $+2$ region of the metal-responsive P_T promoter, suggesting that mercuric ion alters some step leading to the formation of the open complex.

An *in vitro* analysis of the events leading to open complex formation at the bidirectionally transcribed *mer* promoters is complicated by the overlapping structure of those promoters. Open complexes at P_R are detected only in the absence of exogenously added MerR protein. But a large percent of the DNA fragment used to form those complexes remains unbound, giving rise to a background of unprotected nucleotides in footprinting studies. Thus the characterization of those complexes necessitated their separation and isolation from gel retardation assays. Separation by gel retardation and analysis demonstrated that open complexes at the metal-responsive promoter P_T were formed only in the presence of MerR protein and mercuric ion. Intermediate complexes leading to open complex formation are, in general, transient and evasive to experimental characterization. They are often sensitive to competition by heparin, non-specific DNA competitors, high salt, and temperature, and are not stable in the gel separation protocol that has facilitated characterization of the open complex.

The utility of the hydroxyl radical footprinting method is underscored by the footprint of the Hg-MerR-induced open complex. This large assembly includes the multisubunit RNA polymerase holoenzyme (450 kD), Hg-MerR (32 kD), and a several hundred base-pair-long fragment of DNA. The hydroxyl radical protocol provides extensive information about the spatial relationship between these factors. An outline of the footprinting results for the open complex at P_T is shown in the lower portion of Figure 1. Contributions of both Hg-MerR and RNA polymerase have been assigned from footprints of each protein alone. Some ambiguities remain at the interface; this region is marked with crosshatching. This spatial relationship between an allosterically modulated regulatory protein and RNA polymerase is unprecedented. Most transcriptional activator proteins do not bind adjacent to the melted out -10 region but are typically found upstream of the -35 region.

The proximity of the MerR regulatory protein to the melted-out region of the polymerase open complex has mechanistic implications. We are currently exploring two models for the metal-responsive transcriptional activation event (*17*). Initially MerR undergoes a conformation change when Hg(II) binds at the receptor site. This change is manifested as a small but reproducible Hg-induced decrease in the protein/DNA binding constant at room temperature (*2*). Similar results are obtained at 37 °C. Other workers have observed a Hg-induced increase in the protein/DNA interaction, however binding under those conditions could only be observed at 4 °C and may not correlate with physiological properties. MerR isolated from a Gram-positive bacterial strain also shows an Hg(II)-induced decrease in the protein-DNA interaction (*18*). This Hg-induced MerR conformation could promote isomerization to the open complex either a) by altering the contacts of MerR with RNAP through protein-protein interactions, or b) by altering the local structure of the DNA in a protein/DNA mechanism. Analysis of the footprints of the mercuric ion-induced open complex strongly support the latter possibility. These arguments are outlined in (*2*).

In conclusion, it is clear that inorganic reagents can provide detailed structural information concerning the specific interactions of biopolymers in large assemblies. The wide range of properties of coordination complexes suggests that research in this area will continue to provide new insights into biopolymer structure and function.

Literature Cited

1. O'Halloran, T. V. In "Metal Ions in Biological Systems", Sigel, H., Ed., Marcel Dekker, New York, **1989**, Vol. 25, pp 105–142.
2. O'Halloran, T. V. ; Frantz, B.; Shin, M.; Ralston, D.; Wright, J. *Cell* **1989**, *56*, 119–129.
3. Shewchuk, L. M.; Helmann, J. D.; Ross, W.; Park, S. J.; Summers, A. O.; Walsh, C. T. *Biochemistry* **1989**, *28*, 2340–2344.
4. Lund, P. A., Brown, N. L. *J. Mol Biol* **1989**, *205*, 2.
5. Heltzel, A.; Totis, P.; Summers, A. O. In "Metal Ion Homeostasis", Hamer, D. H., Winge, D. R., Eds.; Alan R. Liss, New York, **1989**, pp 427–438.
6. Silver, S.; Misra, T. K. *Ann. Rev. Microbiol.* **1988**, *42*, 717–743.
7. Wee, S.; Neilands, J. B.; Bittner, M. L.; Hemming, B. C.; Haymore, B. L.; Seetharam, R. *Biol. Metals* **1988**, *1*, 62–68.
8. Fürst, P.; Hu, S.; Hackett, R.; Hamer, D. *Cell* **1988**, *55*, 705–717.
9. Aziz, M.; Munro, H. N. *Proc. Natl. Acad. Sci. USA* **1987**, *84*, 8478.
10. O'Halloran, T. , Walsh C. *Science* **1987**, *235*, 211–214.
11. Shewchuk, L. M.; Verdine, G. L.; Walsh, C. T. *Biochemistry* **1989**, *28*, 2331–2339.

12. Tullius, T. D.; Dombroski, B. A.; Churchill, M. E. A.; Kam, L. *Meth. Enzymol.* **1987**, *155*, 537–558.
13. Kierkegaard, K.; Buc, H.; Spassky, A.; Wang, J. C. *Proc. Natl. Acad. Sci. USA* **1983**, 2544–2548.
14. Boroweic, J. A.; Zhang, L.; Sasse-Dwight, S.; Gralla, J. J. *Mol. Biol.* **1987**, *196*, 101–111.
15. Hayatsu, H.; Ukita, T. *Biochem Biophys. Res. Commun.* **1967**, *29*, 556–561.
16. Lund, P. A.; Ford, S. J.; Brown, N. L. *J Gen. Microbiol.* **1986**, *132*, 465–480.
17. Ralston, D.; Frantz, B.; Shin, J.; Wright, J.; O'Halloran, T. V. In "Metal Ion Homeostasis", Hamer, D. H.; Winge, D. R., Eds.; Alan R. Liss, New York, **1989**; pp 407–416.
18. Helmann, J. D.; Wang, Y.; Mahler, I.; Walsh, C. T. *J. Bacteriol.* **1989**, *171*, 222–229.

RECEIVED May 11, 1989

Chapter 8

The Fur Regulon of *Escherichia coli* K-12

Kayoko Nakamura[1], Victor de Lorenzo[2], and J. B. Neilands[3]

Department of Biochemistry, University of California, Berkeley, CA 94720

The binding of ferrous ion to the Fur repressor, the product of the fur *gene, was demonstrated by equilibrium dialysis. The concentration of Fur, in the presence of excess divalent metal activator, required to half-saturate a DNA fragment carrying the promoter of the aerobactin operon was estimated to be about 7 nM. This tight binding plus the previously reported consensus "iron box" sequence of the operator 5'ATAATGATAATCATTAT points to a role for the Fur protein as an important regulator of the siderophore genes in* vivo. *Evidence could not be found for a Fur-Fe(II) catalyzed cleavage of DNA at the operator site of the* fur *gene. The amount of labeled iron absorbed by an E.* coli fur *mutant grown in either minimal or in complex medium containing excess metal ion was not substantially different from that observed with the wild type cell, in spite of enhanced excretion of siderophore by the mutant.*

ALTHOUGH IRON IS GENERALLY REGARDED as a nutritious and essential element, it is not without toxic effects in biology. It is now realized that the element is a major factor in the initiation and catalysis of free radical reactions involved in oxygen-dependent tissue damage (1). Paradoxically, it is the same redox property of iron making it a versatile electron transfer agent in biology that renders it, at the same time, an efficient generator of •OH *via* Fenton-type chemistry. Thus the living cell has a special motivation to regulate with extreme care its total inventory of iron and, in an aerobic milieu, to confine its coordination to molecular species which will minimize the potential for oxidative injury.

Over the past few billion years the surface of the planet has been oxidized by O_2 emanating from the photosynthetic activity of algae and green plants so that the bulk of the environmental iron is now in the higher oxidation state. This is the form of iron that must be solubilized and transported by microorganisms and plants.

[1]Current address: Radiology Department, Keio University, Tokyo, Japan
[2]Current address: Gesellschaft fur Biotechnologische Forschung, Braunschweig, Federal Republic of Germany
[3]Address correspondence to this author.

0097–6156/89/0402–0106$06.00/0
© 1989 American Chemical Society

In 1952, the isolation of a novel iron chelate from the basidio-mycetous fungus *Ustilago sphaerogena* was reported (*2*). The compound, named ferrichrome, seemed unsuited for a role in electron transfer since it bound strongly only trivalent iron. Since *U. sphaerogena* has the physiological quirk of over-synthesis of cytochrome *c*, a role for ferri-chrome in iron assimilation was suspected. This suspicion was strength-ened by the subsequent observation that the synthesis of the ferrichrome ligand could be augmented greatly by culture of the fungus at low iron (*3*). The general formation of ferrichrome type compounds by fungi as well as by Gram negative and positive bacteria served to enhance interest in this line of natural products, now collectively termed siderophores. In all cases, synthesis was shown to be tightly regulated by the available iron supply.

When the focus of research on siderophores switched to the genet-ically more accessible enteric bacteria, it soon became evident that these iron complexes require specialized receptors in order to transit the select-ively permeable outer membrane of the cell. Thus the ferrichrome recep-tor of *Escherichia coli* was equated with the receptor for bacteriophage T1, mutants of the organism displaying both resistance to infection and inability to bind the siderophore (*4*). It appears that siderophore receptors have, in the course of evolution, become "parasitized" by a group of lethal agents, which includes bacteriocins and antibiotics in addition to bacteriophages.

An important advance was registered by McIntosh and Earhart (*5*) who found that iron appears to coordinately repress synthesis of both the endogenous siderophore of *E. coli*, enterobactin, and the outer membrane protein involved in transport of the complexed iron. Obviously, in order to understand the molecular basis of regulation of siderophore synthesis and transport the DNA must be cloned and the systems reconstituted *in vitro*. It is natural that the first cloning experiments in the siderophore series were performed with the genetically approachable *E. coli*. Surpris-ingly, the genetic determinants for enterobactin synthesis and transport of its ferric derivative were found to be spread over more than 20 kb of DNA and to be organized into several transcriptional units (*6, 7*).

In 1979, Williams (*8*) announced the discovery of a novel hydrox-amate type siderophore that was encoded on plasmids, of which pColV-K30 is prototypical, harbored by clinical isolates of *E. coli*. He also showed that the siderophore, subsequently identified as aerobactin (*9*), is an important component of the virulence armamentarium of hospital isolates of *E. coli*. Since the binding of ferric aerobactin had just been found to be the biochemical function of cloacin (*10*), a bacteriocin from *Enterobacter cloaceae*, the acquisition of sensitivity to this agent could be used for selection of cells containing the cloned receptor. In this case the complete regulatory, biosynthetic and specific transport genes were con-fined to only about 8 kb of DNA, a circumstance which facilitated characterization of the operon and determination of the mode of its control by iron (11).

Transcriptional regulation of the aerobactin operon by iron was demonstrated by quantitation of specific mRNA via S1 nuclease protection and by construction of operon and protein fusions with lacZ (12). Considering the regulatory form of iron, the solubility product constant of 10^{-38} for Fe(OH)$_3$ will limit the concentration of free ferric ion to less than 10^{-18} M at biological pH. At the same pH Fe(II) is reported to be soluble to the extent of 100 mM (13). Again, however, it is unlikely that the naked aquo-ion could possess the requisite specificity to stand alone as the regulatory species. Indeed, in 1978 Ernst et al. (14) described a mutation in *Salmonella typhimurium* that led to constitutive expression of all high affinity iron uptake systems in the cell. The lesion was designated *fur* (*ferric uptake regulation*) but was not mapped and has not been studied further in *S. typhimurium*. The *fur* mutation has been replicated in *E. coli* (15), mapped near 16 minutes on the chromosome (16, 17) and the gene cloned and sequenced (18). The 17 kDa Fur protein has been expressed and isolated in the pure state (19).

With the pure Fur protein available it was possible to show that it requires Fe(II) or some other divalent heavy metal ion as co-repressor (20) and that the operator consists of an "iron box" array with the core sequence 5'-ATAATGATAATCATTAT-3' (21). The Fur-divalent metal complex appears to react with the operator as a dimer (20) and engages the DNA via a regular distribution of contacts around the α-helix (22).

While the data collected to date *in vivo* and *in vitro* seem to assure a role for the Fur protein in regulation of expression of siderophores and their outer membrane receptors, it is clear that the *fur* mutation is pleiotropic. The first mutants isolated in Tubingen (15) were acquired by chemical mutagenesis followed by screening for constitutive derepression of a fusion in the gene for the ferrichrome receptor. In Berkeley (17) the corresponding mutants were isolated by screening for constitutive expression, after insertion of Tn5 into the chromosome, of lacZ fused as reporter gene in the aerobactin operon. In spite of the different approaches, the same gene and gene product were obtained. Without exception all *fur* mutants obtained to date display the additional phenotype of failure to grow on certain small, non-fermentable carboxylic acids (23, 24, 25). The mutants are also resistant to manganese (25) and it has recently been reported that they lack SodB, the iron-containing form of superoxide dismutase (26).

Given the above, it is urgently necessary that additional *fur* type mutations be created in bacteria and fungi. In the case of Gram negative species application of the Chrome Azurol S method for detection of siderophore excretion on plates should enable selection of biosynthetic, transport and regulatory mutants (27). In fact, this hope has recently been realized with the symbiotic nitrogen fixer *Rhizobium meliloti* 1021, a genomic region of about 35 kb being sufficient to complement all three types of iron assimilation mutants (*28*).

In spite of considerable published work (*20, 21, 22*), certain important questions regarding the Fur system have thus far not been examined. For instance, does Fur actually bind Fe(II) or is the protein merely "activated" in some way to recognize the DNA? Can an operator fragment be titrated with the Fur-metal ion complex to afford an estimate of the binding affinity? Having bound a protein which contains iron, does the operator DNA then fall victim to oxidative cleavage? Finally, what is the capacity of a *fur* mutant, constitutive in production of siderophore, to accumulate iron from environments containing an excess level of the element? These aspects of the Fur regulon of *E. coli* have been addressed in the present report.

Binding of Fe(II) to the Fur Protein

Previous characterization of the Fur repressor as an Fe(II)-binding protein was based on the finding that strictly anaerobic conditions were required for Fur and iron dependent *in vitro* repression of transcription-translation of a *lacZ* operon fusion (*20*). In addition, several divalent transition metal ions were observed to act as co-repressor in these experiments as well as in footprinting analyses (*21*). Generally, we have employed Mn(II) as *in vitro* activator because of the relative stability of the divalent oxidation state of this ion. Several attempts have been made in this laboratory to demonstrate direct binding of Fe(II) to Fur using flow dialysis and gel permeation chromatographic techniques, but without success. This experience prompted us to resort to the classical procedure of equilibrium dialysis, using sodium ascorbate as reducing agent to maintain the iron in the divalent oxidation state.

A 1.0 mM solution of $FeCl_3$ in dilute HCl was made radioactive by the addition of 0.1 ml of solution containing 1.1 μg $FeCl_3$ with specific activity of 9 μCi/mg. A 17.6 μM solution of Fur (*19*) was prepared in buffer containing 20 mM Tris-HCl (pH 7.5), 1% sodium ascorbate, 0.15 M NaCl and 1 mM $NaHCO_3$. The bicarbonate was added in the event it is required, as in the case of transferrin, for iron binding. In preliminary experiments the survival of excess ascorbate and the stability of Fe(II) in this buffer over a 24 hr period was confirmed, a time span sufficient to achieve the equilibrium state. Exactly 1.0 ml of Fur solution was placed in each of four dialysis bags made from 11.5 mm No. 3 Spectra/Por dialysis tubing, molecular weight cutoff 3500, which had been treated with EDTA and cysteine to remove contaminating metal ions. The bags were placed in each of four separate 50 ml serum bottles which were then filled with buffer and closed with rubber septa. Approximate volumes of 0.5, 0.6, 0.7 and 0.8 ml were injected through the septa and the bottles gently agitated at 23 °C for 24 hr. The radioactivity was then measured on 0.6 ml volumes drawn from inside the bags and from the external solution. The protein concentration inside the bags, as determined by the dye binding method (29), had in all instances become diluted to 12.5 µM from an initial value of 17.6 µM.

The data provided by Table I indicate that Fur is, in fact, an Fe(II) binding protein. Higher levels of iron could not be equilibrated with Fur at this concentration without precipitation of the protein. Assuming a 1:1 complex, the K_D estimated from these data would be in the range of 10 µM.

Table I. Binding of [59]Fe(II) to the Fur Protein Measured by Dialysis (Equilibrium Concentration of [59]Fe(II) vs 12.5 µM Fur)

Unbound outside (µM)	Total inside (µM)	Bound inside (µM)
10.0	11.4	1.4
11.8	14.2	2.4
13.5	17.8	4.3
14.9	20.3	5.4

Binding of Fur to an Operator Fragment

Retardation of mobility on acrylamide gels can be used to quantitate protein-DNA interactions (30). In order to estimate the affinity of Fur for its operator a 250 bp 3' end labelled EcoRI-PvuII restriction fragment carrying the operator sequences of an iucA'-'lacZ fusion of pCON6 was incubated with increasing concentrations of repressor and analyzed on 4-5% acrylamide gel. The conditions used were 20 °C, 20 mM Bis-Tris/borate buffer (pH 7.5), 1 mM $MgCl_2$, 100 µM $MnCl_2$ and 40 mM KCl. As may be seen from Figure 1, half binding was observed at a concentration of Fur of approximately 7 nM. This value is in good agreement with the recently published figure of 5 nM for a 156 bp Sau3A fragment spanning the major and minor promoter elements of the aerobactin operon, as analyzed in 5.25% polyacrylamide (19).

The Fur-Fe(II) Complex Does Not Cleave DNA

Certain natural and synthetic iron binding compounds are very effective agents for oxidative scission of DNA (31). It was hence of interest to determine if this property, which could pose a hazard to the cell, could be associated with the Fur protein. Experiments were initiated to ascertain if

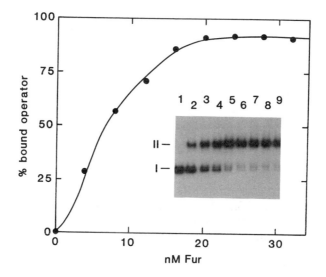

Figure 1. Concentration dependence of Fur-operator binding. A 3'-end labeled EcoRI-PvuII (22) fragment carrying the operator sequences of the aerobactin operon in 250 bp was incubated with increasing concentrations of Fur repressor and analyzed by gel electrophoresis. The nine lanes shown on the inset correspond to the respective points on the graph. The ratios of bound (II) to unbound (I) operator were obtained by densitometry.

Fur-Fe^{2+} complexes are capable of synthesis of radicals in sufficiently close proximity to the DNA to afford sequence-specific cleavage near the site of binding. We have previously observed a few over-exposed bases in •OH footprinting analysis (32) of the aerobactin promoter (22). In the present work the promoter of the *fur* gene rather than the promoter of the aerobactin operon was used since the former has a single Fur-binding site (33).

A 250 bp *Eco*RI-*Dra*I fragment carrying the promoter of the *fur* gene (33) was 3' labelled at the *Eco*RI extremity by use of reverse transcriptase and ^{32}P-dATP+dTTP. The labeled fragment was diluted with 100 μl of footprinting buffer (devoid of EDTA). The experiment was designed to include controls with Fur and DNase I in the presence of two redox active metal ions, Fe(II) and Mn(II), and a non-redox active metal, Cd(II), all with and without additions of Fur, dithiothreitol or ascorbate.

Analysis of these various mixtures on sequencing gels showed that the few especially intense bands are also present in the untreated fragment and hence cannot arise from a specific Fur-metal ion catalyzed cleavage.

Uptake of ^{59}Fe by Escherichia coli BN402 and BN4020 in Various Media

E. coli strains BN402 F⁻ *thr*1 *thi*1 *his*4 *arg*E3 *lac*DU169 *gal*K2 *ara*14 *xyl*15 *mtl*1 *str*31 *tsx*? *sup*?, which is *fur*⁺, and BN4020, a *fur*1::Tn5 derivative of BN402, have been described previously (17). Both strains have the ability to synthesize enterobactin and to transport its ferric complex. When cultured in minimal medium M9 (34) the required amino acids were added at a level of 40 μg/ml. For strain 4020 kanamycin was added at a concentration of 25 μg/ml. An inoculum of 0.2 ml from an overnight culture of LB was added to 12 ml of growth medium contained in 29 × 115 mM test tubes. Following introduction of labeled element the tubes were shaken at 37 °C. At the indicated intervals of time the absorbancy was measured at 550 nm and the cell mass estimated on the assumption that an A$_{550}$ of 1.0 is equal to 0.38 mg dry weight/ml (35). Simultaneously a 1.0 ml volume was filtered through a membrane with pore size 0.22 μm. The filter was washed with two, 1 ml aliquots of sterile medium and the radioactivity of the membrane was then measured in a well counter.

Inspection of the data in Table II suggests that there is little difference in the amount of ^{59}Fe taken up by the two strains growing in minimal medium containing excess iron even though an Arnow (36) assay showed over ten times as much catechol present after 5.3 hours in the *fur* culture. In this minimal medium the endogenous iron concentration may be assumed to be about 10% of the added level of 10 μM.

Table II. Uptake of ^{59}Fe by *Escherichia coli* Strains BN402 and BN4020
Cultured in Minimal Medium M9 Containing 10 μM $FeSO_4$

Incubation Period (hr)	Cell Density (A$_{550}$ nm)		[59]Fe in Cells (nmoles/mg dry wt)	
	402	4020	402	4020
1.4	0.140	0.102	35.8	42.9
2.8	0.482	0.296	23.7	24.7
3.8	0.938	0.594	24.5	21.1
5.3	1.190	1.170	20.0	17.6

When this experiment was repeated in medium LB (*34*), which contains yeast extract and tryptone and hence probably has an iron content of about 30 μM, uptake of the label was severely reduced (Table III). Again, however, the *fur* strain 4020 enjoyed no significant advantage as regards accumulation of iron.

Table III. Uptake of ^{59}Fe by *Escherichia coli* Strains BN402 and BN4020
Cultured in Complex Medium LB Containing 10 μM $FeSO_4$

Incubation Period (hr)	Cell Density (A$_{550}$ nm)		[59]Fe in Cells (nmoles/mg dry wt)	
	402	4020	402	4020
1	0.170	0.148	1.16	0.82
2	0.528	0.418	1.39	0.63
3	1.050	0.936	2.34	1.71
4	1.564	1.440	2.55	1.87

To confirm that the form of iron supplied is not responsible for these data, strains 402 and 4020 were compared in LB medium supplemented with a trace of ^{59}Fe ferric citrate. Once more, as shown in Table IV, no enhanced absorption of label could be assigned to the *fur* strain.

Table IV. Uptake of ^{59}Fe by *Escherichia coli* Strains BN402 and BN4020 Cultured in Complex Medium LB Containing Ferric Citrate

Incubation Period (hr)	Cell Density (A$_{550}$ nm)		^{59}Fe in Cells (nmoles/mg dry wt)	
	402	4020	402	4020
1	0.160	0.140	1.61	2.09
2	0.540	0.434	2.00	2.11
3	1.048	0.922	3.82	2.76
4	1.528	1.418	4.16	3.11

Discussion

Deletion of high-affinity, siderophore-mediated microbial iron assimilation does not constitute a lethal mutation, although it may affect virulence of species pathogenic for man (8, 37), animals (38) and plants (39) and deny growth in laboratory media containing iron complexing agents. Thus mutants defective in siderophore synthesis or transport can be cultivated on simple media and complemented with genomic fragments to test for restoration of the wild phenotype. Cloning of these systems should be relatively straightforward. The data assembled in Table V indicate that some progress in this direction has been achieved. However, in regard to mechanism, only in the case of the aerobactin operon of pColV-K30 has the incisive •OH footprinting method (32) been applied in order to reveal details of the contacts of repressor with DNA (22). Transcriptional regulation mediated by the Fur repressor-ferrous iron complex appears to be a major, although possibly not exclusive, control element. Thus we have yet to examine the possibility that iron may also affect the stability of the message and there remains a need to search for cruciform type structures in the aerobactin operon.

Table V. Cloning Experiments in the Siderophore Series

Siderophore	Chemical Type	Source	Reference
Enterobactin	Catechol	*Escherichia coli*	*(6, 7)*
Aerobactin	Hydroxamate; α-hydroxy acid	*Escherichia coli*	*(11)*
Pseudobactin	Catechol; hydroxamate α-hydroxy acid	*Pseudomonas* sp.	*(40)*
Anguibactin	Catechol;?	*Vibrio anguilarum*	*(38)*
Ferrioxamine B	Hydroxamate	*Streptomyces pilosus*	*(41)*
Rhizobactin 1021	?	*Rhizobium meliloti* 1021	*(28)*
Chrysobactin	Catechol	*Erwinia chrysanthemi*	*(39)*

In addition to the variety of phenotypes already listed for the *fur* mutation in *E. coli* K-12, we should remark that the presence of a functional CAP site in the upstream region of the *fur* gene itself gears expression of the repressor to the cAMP system of the cell. The Fur protein has been identified by Dr. Liz Hutton (personal communication) on the two-dimensional protein map of *E. coli* and has been assigned the designation G15.8. It should now be possible to monitor its levels under different physiological conditions. This is the type of study which, at the present time, can only be performed in *E. coli.*

Regardless of the medium used we have observed a slight but significantly lower growth rate by our *fur* mutant, which is likely a reflection of the pleiotropic nature of the mutation. In spite of its enhanced level of expression of the enterobactin system the mutant did not, however, accumulate significantly more iron through its exponential and stationary growth phases when contrasted with the parent strain. The iron content of the dry cell mass recorded in Table II is certainly a minimum value since the adventitious iron was not labeled and we have no means to estimate its possible preferential use over the ferrous sulfate supplied. Nonetheless the values reported in Table II are comparable to those found in the literature. An average figure of about 20 nmoles/mg dry weight has been given for *E. coli* growing in minimal medium supplemented with iron to a level of 5 μM (5). At these higher levels there is a possibility that some of the iron has merely adhered to the cell envelope.

It must be stressed that while enterobactin and its outer membrane receptor are constitutively expressed in the *fur* mutants, it is less certain that the several other genetic loci needed for utilization of complexed iron, such as *ton*B, are similarly induced. Indeed, Hantke (15) found no low iron-enhanced expression of *lacZ* fused in gene *ton*B, the latter a function required for transport of all ferric siderophores. Alternatively, failure of the *fur* strain to accumulate excess iron may be related, in some manner not yet understood, to the pleiotropic character of the mutation.

In enteric bacteria the enterobactin system appears to operate as a device to satisfy the minimal essential iron requirement of the cell when the element is in short supply. Thus the *ent* mutants of *S. typhimurium* blocked between chorismate and enterobactin have similar generation times as the wild type in both complex and minimal media. However, the *ent* mutants suffer severe iron deficiency when citrate, the iron complex of which cannot be transported by *S. typhimurium*, is added to the minimal medium (42). Similarly, Rosenberg and Young (43) concluded that in *E. coli* enterobactin is "only required when the concentration of iron in the growth medium is very low."

By the use of equilibrium dialysis we have been able to show direct binding of Fe(II) to the Fur protein in the presence of ascorbate, a reducing agent which makes no detectable complex with ferrous ion (44). The data in Table I are not sufficiently extensive to allow a Scatchard analysis of the binding constant and determination of the number of ferrous ions bound per mole of Fur. It will be necessary to extend this study to a broad range of Fur/Fe(II) ratios and to experiment with a variety of additional techniques to maintain the iron in the reduced state. The affinity of Fe(II) for Fur, however, is sufficient to conclude that the metal ion is bound to the protein in the course of DNase I footprinting analysis of the aerobactin operator (21). A more precise measure of the affinity of divalent metal ions for Fur and their sites of binding in the protein are problems currently under study by NMR techniques in the laboratory of R. J. P. Williams at Oxford.

Similar comments apply to estimation of the precise affinity of Fur, in the presence of excess divalent metal ion as activator, for the operator sequences. The binding curve is compatible with a second-order dependence on the Fur concentration, which would be expected if Fur dimers were the species active in binding DNA and Fur monomers were the predominant unbound forms. However, we are not certain of the oligomeric form of either bound or free Fur and the curve illustrated in Figure 1 could be the result of a more complicated equilibria. The Fur protein, although relatively small (17 kDa), produces quite a substantial shift in mobility and it is possible that several copies of both protein and metal ion are present in the operator complex. The stoichiometry of this complex is presently under investigation.

Our final remark concerns the failure of Fur-Fe(II) mixtures to cleave the DNA at the operator site. In order for this to constitute a danger to the cell, O_2 would have to contact the complex *in vivo* and there is every reason to believe that the interior of *E. coli* is a highly anaerobic environment. In addition, it is the usual observation that iron bound to protein, rather than complexed in chelates such as EDTA, is not prone to generate oxidizing radicals (1). However, a potential problem of iron toxicity at the

DNA level could be circumvented *via* a positive regulatory model wherein an activator protein binds the DNA only at low iron. In *Pseudomonas aeruginosa* the structural gene for the ToxA endotoxin is iron-controlled by means of a separate gene product acting as a positive regulator of transcription (45). It is tempting to speculate that the *Pseudomonas*, which are obligate aerobes, may have adopted this particular mode of regulation for protection of DNA sequences necessarily exposed to both iron and oxygen.

Acknowledgments

This research was supported in part by grants AI04156, PC78-12198 and CRCR-1-1633 from the NIH, USDA and NSF, respectively. The authors are indebted to Dr. Fabio Giovannini for the data recorded in Figure 1.

Literature Cited

1. Halliwell, B. (ed), in "Oxygen Radicals and Tissue Injury", Federation of American Societies for Experimental Biology, Bethesda, MD, **1988**.
2. Neilands, J. B. *J. Am. Chem. Soc.*, **1952**, *74*, 4846.
3. Garibaldi, J. A.; Neilands, J. B. *Nature*, **1956**, *177*, 526.
4. Wayne, R.; Neilands, J. B. *J. Bacteriol.*, **1975**, *121*, 497.
5. McIntosh, M. A.; Earhart, C. F. *J. Bacteriol.*, **1977**, *131*, 331.
6. Laird, A. J.; Ribbons, D. W.; Woodrow, G. C.; Young, I. G. *Gene*, **1980**, *11*, 347.
7. Laird, A. J.; Young, I. G. *Gene*, **1980**, *11*, 359.
8. Williams, P. H. *Infect. Immun.*, **1979**, *26*, 925.
9. Warner, P. J.; Williams, P. H.; Bindereif, A.; Neilands, J. B. *Infect. Immun.*, **1981**, *33*, 540.
10. van Tiel-Menkveld, G. J.; Mentjox-Vercuurt, M.; Oudega, B.; de Graff, F. K. *J. Bacteriol.*, **1982**, *150*, 490.
11. Bindereif, A.; Neilands, J. B. *J. Bacteriol.*, **1983**, *153*, 1111.
12. Bindereif, A.; Neilands, J. B. *J. Bacteriol.*, **1985**, *162*, 1039.
13. Hay, R. W., "Bio-Inorganic Chemistry", Ellis Horwood, Ltd., Chichester, U.K., **1984**.
14. Ernst, J. F.; Bennett, R. L.; Rothfield, L. I. *J. Bacteriol.*, **1978**, *135*, 928.
15. Hantke, K. *Mol. Gen. Genet.*, **1981**, *182*, 288.
16. Hantke, K. *Mol. Gen. Genet.*, **1984**, *197*, 337.
17. Bagg, A.; Neilands, J. B. *J. Bacteriol.*, **1985**, *161*, 450.
18. Schaffer, S.; Hantke, K.; Braun, V. *Mol. Gen. Genet.*, **1985**, *201*, 204.
19. Wee, S.; Neilands, J. B. *Biol. of Metals*, **1988**, *1*, 62.

20. Bagg, A.; Neilands, J. B. *Biochemsitry*, **1987**, *26*, 5471.
21. de Lorenzo, V.; Wee, S.; Herrero, M.; Neilands, J. B. *J. Bacteriol.*, **1987**, *169*, 2624.
22. de Lorenzo, V.; Giovannini, F.; Herrero, M.; Neilands, J. B. *J. Mol. Biol.*, **1988**, *204*, 875.
23. Bagg, A.; Neilands, J. B. *Bacteriol. Rev.*, **1987**, *51*, 509.
24. Neilands, J. B.; Konopka, K.; Schwyn, B.; Coy, M.; Francis, R. T.; Paw, B. H.; Bagg, A., in "Iron Transport in Microbes, Plants and Animals", G. Winkelmann, D. van der Helm and J. B. Neilands (eds), VCH Publishers, Mannehim, FRG, **1987**.
25. Hantke, K. *Mol. Gen. Genet.*, **1987**, *210*, 135.
26. Niederhoffer, E. C.; Naranjo, C.; Fee, J. A. *J. Cell Biochem.*, **1988**, Suppl. 12D, 340.
27. Schwyn, B.; Neilands, J. B. *Anal. Biochem.*, **1987**, *160*, 47.
28. Gill, P.; Neilands, J. B., manuscript in preparation.
29. Bradford, M. M. *Anal. Biochem.*, **1976**, *72*, 248.
30. Wu, H.; Crothers, D. *Nature*, **1984**, *308*, 509.
31. Griffin, J. H.; Dervan, P. B. *J. Am. Chem. Soc.*, **1987**, *109*, 6840.
32. Tullius, T.; Dombroski, B. *Proc. Natl. Acad. Sci. USA*, **1986**, *83*, 5469.
33. de Lorenzo, V.; Herrero, M.; Giovannini, F.; Neilands, J. B. *Eur. J. Biochem.*, **1988**, *173*, 537.
34. Miller, J. H. in "Experiments in Molecular Genetics", Cold Spring Harbor Laboratory, Cold Spring Harbor, New York, NY, **1972**.
35. Hartmann, A.; Braun, V. *Arch. Microbiol.*, **1981**, *130*, 353.
36. Arnow, L. E. *J. Biol. Chem.*, **1937**, *118*, 531.
37. Bullen, J. J.; Griffiths, E. (eds), in "Iron and Infection", Wiley, Chichester, U.K., **1987**.
38. Crosa, J. H., in "Iron Transport in Microbes, Plants and Animals", G. Winkelmann, D. van der Helm and J. B. Neilands (eds), VCH Publishers, Mannheim, FRG, **1987**.
39. Enard, C.; Diolez, A.; Expert, D. *J. Bacteriol.*, **1988**, *170*, 2419.
40. Moores, J. C.; Magazin, M.; Ditta, G. S.; Leong, J. *J. Bacteriol.*, **1984**, *157*, 53.
41. Schupp, T.; Toupet, C.; Divers, M. *Gene*, **1988**, *64*, 179.
42. Pollack, J. R.; Ames, B. N.; Neilands, J. B. *J. Bacteriol.*, **1970**, *104*, 635.
43. Rosenberg, H.; Young, I. G., in "Microbial Iron Metabolism", J. B. Neilands (ed), Academic Press, New York, **1974**.
44. Hamed, M. Y.; Keypour, H.; Silver, J.; Wilson, M. T. *Inorganica Chimica Acta*, **1988**, *152*, 227.
45. Wozniak, D. J.; Cram, D. C.; Daniels, C. J.; Galloway, D. R. *Nucleic Acids Res.*, **1987**, *15*, 2123.

RECEIVED May 11, 1989

Chapter 9

Platinum Anticancer Drug Binding to Oligonucleotide Models of DNA

^{31}P NMR Spectroscopic Investigations

Luigi G. Marzilli[1], Tammy Page Kline[1], David Live[1], and Gerald Zon[2]

[1]Department of Chemistry, Emory University, Atlanta, GA 30322
[2]Applied Biosystems, 850 Lincoln Centre Drive, Foster City, CA 94404

^{31}P NMR spectroscopic investigations on the binding of anticancer drugs to DNA are useful when sufficient structural perturbations are introduced to yield changes in shifts of ^{31}P NMR signals by ~0.5 ppm. Although such large shifts in ^{31}P signals require significant changes in torsion and/or bond angles at the phosphodiester linkage, drugs that are intercalators or that form covalent links to the DNA appear to cause these types of structural changes. Chemical methods for assigning ^{31}P signals are reviewed and compared to methods involving magnetization transfer, e.g., HMQC (heteronuclear multiquantum coherence), selective reverse chemical shift correlation (SRCSC), and heteronuclear 2D J coupling experiments. Surveys of the effects of Pt compounds with the characteristic structural features cis-PtA_2X_2, where A = amine with at least one NH group and X = leaving ligand, have revealed that significant downfield shifted signal intensity occurs only when the target DNA has adjacent G residues. In contrast, downfield signals have not been observed with the inactive compounds, [Pt(diethylenetriamine)Cl]Cl or trans-Pt(NH$_3$)$_2$Cl$_2$ (an inactive isomer of the clinically used drug cis-Pt(NH$_3$)$_2$Cl$_2$). In general, when such downfield signals have been unambiguously identified, they clearly arise from the phosphodiester group between the two G's. The characteristic features of such spectra will be discussed, and then the issue will be raised regarding the possibility that other phosphate groups can give such downfield signals.

NMR SPECTROSCOPY HAS PLAYED AN IMPORTANT ROLE in deducing the nature of the interaction of anticancer drugs with DNA (1), the most common molecular target of such drugs. Recently, the availability of sufficient quantities of synthetic oligonucleotides has greatly increased the

0097–6156/89/0402–0119$07.50/0
© 1989 American Chemical Society

number of NMR studies. The typical nucleus of choice for NMR spectro-scopic investigation of drug binding to oligonucleotides is [1]H. The high γ value of this nucleus makes it the most easily observed nucleus commonly present in both DNA and anticancer drugs. Furthermore, modern homonuclear 2D NMR methods (COSY and NOESY) have made spectral assignments relatively routine, particularly for standard struc-tural forms of DNA, e.g., right-handed A or B form DNA or left-handed Z DNA (2). For example, with 2D COSY spectroscopy it is possible to identify spin coupled protons (H5, H6 of cytosine bases; H1', H2'/2", H3', and H4' of the deoxyribose). By 2D NOESY, internucleotide NOE's can be observed between base protons (H8 of adenine or guanine, H6 of thymine or cyto-sine) and sugar protons (particularly H1' and H2' and/or H2") of the 5'-nucleotide. There are also internucleotide NOEs between the 5'AH8 and the adjacent 3'T methyl group. (2, 3). By combining these 2D connectivi-ties, the ^{1}H NMR signals can be readily assigned for standard DNA con-formations. This methodology has proved to be a powerful tool in assess-ing drug-DNA interactions. Fairly detailed conformational/structural in-formation can be obtained by quantitation of NOE crosspeak intensities, which are related to the inverse sixth power of the distance between protons (3–12).

Despite the power and success of these approaches, there are, how-ever, some significant limitations of the use of ^{1}H NMR spectroscopy. First, except under favorable conditions of high temperature (particularly for synthetic polymers), ^{1}H NMR signals are broad for polymeric DNA. Therefore, it may not be possible to correlate the spectral changes in the polymer with those of the oligonucleotides. This situation may lead to uncertainty concerning the relevance of studies on drug-oligonucleotide adducts to drug-DNA binding. Second, assignments may be complicated by the loss of internucleotide connectivities brought on by the structural change on drug binding. These sequential connectivities depend greatly on base-to-sugar NOE's (2, 3), and a significant structural change could eliminate these connectivities. However, this problem can be overcome if protons in the drug act as effective "conduits" of connectivities (13). Third, there can be difficulty in interpreting the multitude of cross peaks in a typical 2D NMR spectrum if more than one binding mode occurs simultaneously. In 1D spectroscopy, multiple species are easily detected and peaks attributable to each are identified by relative intensity (13–15). However, the intensities of cross peaks are a function of several factors such as the efficiency of transfer of magnetization, the relative abundance of the species, etc. Thus, a detailed ^{1}H NMR analysis of a drug-oligonu-cleotide complex is usually feasible only when the model can be designed to be both representative of the principal type of interaction and to have only one species in solution. Fourth, even at 500 or 600 MHz, signal over-lap can be significant, particularly the H4' and H5'/5" signals, but also for the H3' signals.

The problems mentioned above in some cases also apply to the spectroscopy of less sensitive nuclei (^{31}P, ^{13}C, and ^{15}N) (*16–23*) which also occur in DNA and/or the drugs. Furthermore, other problems in addition to lower sensitivity also arise. Here, we shall consider ^{31}P NMR spectroscopy only.

The ^{31}P NMR signals of uncomplexed DNA fall in a relatively narrow range since, prior to the addition of a drug, the bond and torsion angles of all the phosphodiester linkages in DNA are essentially identical (*24, 25*). The relaxation of NMR signals of nuclei such as ^{31}P can be affected by chemical shift anisotropy that will tend to broaden signals from polymeric materials, particularly at high field. However, segmental motion of the sugar phosphate backbone compensates for this and ^{31}P signals of DNA do not broaden severely with increasing field (*26*). With the addition of a drug to DNA, significant changes in torsion and/or bond angles at ^{31}P can then lead to observable new signals well resolved from the main body of resonances arising from regions with "normal" structure. These shifted signals can be viewed as a "fingerprint" of the type of interaction of a drug with a DNA polymer, and an appropriate ^{31}P signal from a drug-oligomer complex can be taken as an indication that the oligonucleotide complex is a good model for the drug binding site.

In order to establish the locations of the phosphate groups with shifted ^{31}P signals, it is generally necessary to assign all of the phosphorus atoms in the molecule. This can be complicated by the overlap of ^{31}P resonances in the region of normal signals. An approach to making assignments is to correlate the ^{31}P's to the sugar protons, particularly H3', to which they are coupled three bonds away. The added dispersion of heteronuclear 2D methods with the aid of previously assigned proton signals could allow assignments of the ^{31}P NMR signals (*27–39*).

Conventional methods for heteronuclear 2D experiments employ direct observation of the heteronucleus, e.g. phosphorus. While technically easier to carry out than proton-detected heteronuclear 2D, the former experiments are less sensitive and offer poorer digital resolution in the *proton* dimension where it is generally needed most. There are several options for proton-detected ^{1}H-^{31}P 2D experiments. Proton-detected heteronuclear multiquantum coherence (HMQC) experiments have been used successfully in making ^{1}H-^{31}P correlations (*40*). Although these experiments work well for nuclei like ^{13}C and ^{15}N where the interaction is typically mediated by a one-bond coupling to the proton from the heteronucleus that is much larger than other homonuclear couplings (*41*), in cases where homo- and heteronuclear couplings are comparable, as for ^{31}P, the efficiency of such experiments is diminished (*30*). Another limitation of these experiments in this application is that it is difficult to get phased 2D spectra, and therefore absolute value mode spectra are obtained with lower resolution.

An alternative to the HMQC experiment is to use a selective reverse chemical shift correlation (SRCSC) experiment (*29*) which can be phased.

In principle, this experiment is less sensitive since the lower ^{31}P polarization is being transferred to the protons; however, since signal is not lost through the generation of higher order multiquantum coherences, the results are superior to HMQC (29). A modified version of SRCSC includes a selective 180° pulse in the evolution period that effectively decouples the H3' protons from other protons, yielding a 2D map that contains a $^1H–^{31}P$ shift correlation with the cross-peaks split only by the vicinal H3'–C–O–P coupling (30). The selection for this single coupling constant provides significant additional information on bond angles in addition to aiding in assignments. An important advantage of the $^1H–^{31}P$ SRCSC experiment is the added dispersion in 2D that makes it possible to resolve signals from overlapping H3' protons. Another experiment that should provide more accurate vicinal coupling constants, and also aids in identifying H3' protons with very weak coupling to ^{31}P nuclei (when coupling is small these peaks are difficult to see in the COSY-like SRCSC experiment), is a proton-detected heteronuclear 2D J experiment, again depending on selective excitation of the H3' signals (30). We have recently found the latter two experiments to be particularly useful in Pt–drug DNA studies in assigning resonances and identifying unusual structures.

Chemical modification of the phosphates by ^{17}O labeling methods (31, 42–45) has been used with a considerable amount of success recently for assignments of ^{31}P signals of oligonucleotides as well as their reaction products with various drugs. An oligodeoxyribonucleotide is synthesized *via* the phosphoramidite method (46) in which the phosphite ester is oxidized in the presence of water which is *ca.* 40% $H_2^{17}O$, 40% $H_2^{18}O$, and 20% $H_2^{16}O$. The ^{31}P signal of the labeled phosphate group is affected differently by each of the three oxygen isotopes. In the case of ^{17}O, the $^{31}P(^{17}O)$ signal is broadened by the quadrupolar ^{17}O nucleus. An isotope shift effect is observed for the ^{18}O species and both $^{31}P(^{16}O)$ and $^{31}P(^{18}O)$ signals are observed with *ca.* 0.03 ppm separation (47–49). Thus, the ^{31}P signal due to the labeled phosphate group exhibits *ca.* 40% reduction in intensity as well as a "doublet" appearance. This procedure unambiguously assigns the signal (31, 42–45).

Joseph and Bolton (42) compared spectra of poly(I)•poly(C) unlabeled and labeled with ^{17}O in the nonbridging positions of poly(I) and were able to assign the upfield signal to poly(C). Petersheim et al. (43) were able to assign the ^{31}P resonances of d(CGCG) before and after addition of Actinomycin D (Act D). In a similar manner, all ^{31}P signals of d(GGAATTCC) and d(CGCGAATTCGCG) have been assigned (44, 45). Gorenstein et al. (31) used a combination of ^{17}O-labeling and a 2D $^{31}P–^1H$ chemical shift correlated NMR spectral technique to assign the ^{31}P spectrum of d(AGCT) and its Act D complex.

The problem with 2D ^1H NMR spectroscopy in the presence of multiple species does not apply to typical 1D ^{31}P NMR spectra. The presence of several drug binding sites can be easily identified if, as is often the case, large differential shifts of phosphodiester groups are induced at the drug binding sites (50). In favorable cases the [17]O labeling method allows signal assignment of affected phosphate groups in multiple coexisting species (50). In addition, the greater shift range of ^{31}P NMR spectra could allow identification of species interconverting too rapidly to allow distinction by ^{1}H NMR spectroscopy.

Recent examples of the success of ^{31}P NMR spectroscopy in identifying the signals of systems in which multiple products are formed and in providing structural or binding site information include studies with Act D (50) and *meso*-tetrakis(4-N-methylpyridiniumyl)porphyrin [TMpyP(4)] (51). The ^{31}P NMR spectrum of the oligonucleotide d(TGCGCA) was assigned by the [17]O method (50). On addition of Act D, two species were clearly evident at the ratio of one drug per duplex, whereas at the ratio of two drugs per duplex, only one species with five ^{31}P signals was present. On this basis, the latter adduct must have C_2 symmetry, and two downfield signals were assigned to the two GpC sites, consistent with intercalation of the phenoxazone ring of the drug at both GpC sites. At the ratio of one drug per duplex, the signals could be identified and were consistent with two products with the Act D bound off the C_2 center at the GpC sites. The phenoxazone ring is unsymmetrical, a feature that leads to two adducts. Recent 2D ^{1}H NMR studies confirm these conclusions (13).

In contrast, a longer self-complementary oligonucleotide containing the same central sequence used in our Act D studies, namely d(TATA-TGCGCATATA), exhibits one sharp very downfield ^{31}P NMR signal at –0.9 ppm on treatment with TMpyP(4) (51). A similar unusually shifted signal was observed on treating the synthetic polymer poly[d(G-C)]•poly[d(G-C)] with TMpyP(4). The signal was broader, however, as expected for a polymer. [17]O-labeling clearly established the central phosphate group CpG as the porphyrin binding site.

In our view, ^{1}H and ^{31}P NMR spectroscopies are highly complementary, each exhibiting its greatest strengths where the other is weakest. Thus, large drug-induced distortions of DNA structure can interrupt the 2D NMR connectivities so important in the ^{1}H NMR method. However, exactly these distortions induce the largest ^{31}P NMR signal shifts in cases in which [17]O methods as well as heteronuclear [1]H–[31]P methods are most powerful. Likewise, the [1]H–[31]P coupling can permit 2D experiments that effectively disperse the ^{1}H NMR signals, allowing more facile identification of the ^{1}H assignments. When multiple species are present, these may be difficult to characterize by ^{1}H NMR spectroscopy, whereas this situation is often clear with ^{31}P NMR spectroscopy. Sequential signal assignment is particularly difficult in such cases by ^{1}H 2D NMR spectroscopy, but it can be accomplished readily for ^{31}P NMR spectroscopy by [17]O labeling methods and [1]H–[31]P chemical shift correlation experiments. The ^{31}P assignments can be related to ^{1}H by 2D methods mentioned above. In contrast, ^{31}P NMR spectroscopy is not particularly informative in normal, typically B form, regions of DNA and, here, methods for signal assignment are strongest for ^{1}H NMR spectroscopy. Finally, the greatest utility of ^{31}P NMR spectroscopy is for large molecules, whereas ^{1}H NMR spectroscopy is most suited to smaller molecules.

^{31}P NMR spectroscopy is unlikely to displace ^{1}H NMR spectroscopy as the method of choice for studying anticancer drug-DNA complexes for a number of reasons in addition to those just mentioned. In particular, outside binders, which usually bind in DNA grooves, cause minor shifts in ^{31}P NMR signals (52–56). These shifts are often slightly upfield. Although this insensitivity of ^{31}P NMR signals to outside binding by anticancer drugs may be considered disadvantageous, it nevertheless provides strong evidence for non-intercalative binding. Specific interactions between groove binders and the DNA may be readily detected by ^{13}C NMR spectroscopy, but this methodology has received relatively little attention in the study of oligonucleotide models.

Applications to Pt Drug Binding to DNA

Background. Strong evidence exists implicating DNA as the principal target for cis-Pt(NH$_3$)$_2$Cl$_2$ in biological systems (1). Among the purine and pyrimidine bases, Pt seems to favor guanine as a reactant and, in particular, GC rich regions of polynucleotides seem to be preferential targets (1, 57–60). Chromatographic analyses of the enzymatic digestion products of salmon sperm DNA treated with cis-Pt(NH$_3$)$_2$Cl$_2$ (61, 62) revealed that $ca.$ 85% of the bound Pt species were from an intrastrand type of crosslink; $ca.$ 50–60% of these intrastrand products were due to Pt binding to adjacent G residues. Since the effectiveness of a Pt anti-tumor agent may arise from its ability to disrupt the DNA so that excision-repair and DNA-replication enzymes are inhibited (1, 57, 58), the influence of Pt drugs on the ^{31}P spectrum of DNA and related systems has been investigated by several research groups.

Initial studies of the effects of Pt anti-cancer agents on the ^{31}P NMR spectrum of DNA revealed a new peak suggestive of a large conformational change in part of the DNA (63). Reaction of the active anti-tumor agents Pt(en)Cl$_2$, cis-Pt(NH$_3$)$_2$Cl$_2$, and Pt(cyclohexanediamine)Cl$_2$ with salmon sperm and calf thymus DNA [as well as nucleosomes with Pt(en)Cl$_2$] introduces a new peak centered at $ca.$ 1.2 ppm downfield from the untreated DNA signal (64). The inactive compounds $trans$-Pt(NH$_3$)$_2$Cl$_2$ and [Pt(diethylenetriamine)Cl]Cl produced no such new signal. Den Hartog et al. (65) observed a similar downfield signal upon reaction of salmon sperm DNA with cis-Pt(NH$_3$)$_2$Cl$_2$. The downfield resonance is most probably a consequence of some structural distortion in the DNA induced by the active compounds, but not by the inactive ones. On treatment with drug, the nonalternating polymer poly(dG)•poly(dC), but not alternating G–C or any A•T polymers, exhibited a similar spectral change (64). This finding suggested that the structural change in the DNA responsible for the spectral change arises primarily from reaction of Pt anti-cancer agents with adjacent G residues.

Similar studies were conducted with the polyribonucleotide poly(I)•poly(C) (66). The ^{31}P NMR spectrum of poly(I)•poly(C) shows two fairly well-resolved signals at −4.03 and −4.50 ppm assigned to the poly(I) and the poly(C) strands, respectively. Upon reaction with Pt(en)Cl$_2$ at r = 0.2 (r = ratio of Pt to DNA phosphate), two signals of roughly equal intensity at −3.3 and −3.7 ppm and a sharp signal at −4.2 ppm were observed. The two downfield signals were attributed to the phosphate groups between the platinated I residues and the phosphate group between complexes, respectively. The sharp signal corresponds closely to that of single-stranded poly(C), unplatinated. Similar results were obtained with other active anti-tumor Pt compounds. Reaction with inactive compounds merely broadened the ^{31}P signal.

A survey of the effects of Pt compounds on the ^{31}P spectra of a variety of self-complementary oligodeoxyribonucleotides was conducted (67). A non-transient downfield signal was observed at *ca.* −3.0 ppm for each *cis*-Pt(NH$_3$)$_2$Cl$_2$- and Pt(en)Cl$_2$-treated oligonucleotide which contained an NGG moiety. In some cases two downfield signals were observed. No signals were observed outside the normal shift region for any of the oligonucleotides treated with *trans*-Pt(NH$_3$)$_2$Cl$_2$.

1D Decoupling Experiment. As mentioned previously, there are several methods available to assist in assignments of ^{31}P signals. The assignment method utilized by den Hartog et al. (65) in the study of the oligonucleotide d(TCTCGGTCTC)•*cis*-Pt(NH$_3$)$_2$, 1, was observation of the ^{1}H spectrum under simultaneous selective irradiation of a ^{31}P resonance (27). The numbering scheme of 1 is shown below:

```
d(T C T C G G T C T C)•cis-Pt(NH3)2
   1 2 3 4 5 6 7 8 9 10
```

In the ^{31}P NMR spectrum of 1, four signals are shifted outside the normal range: two downfield and two upfield. By selective irradiation at a particular ^{31}P resonance the ^{31}P–^{1}H coupling of the H3' and H5'/5" protons is eliminated. Due to overlapping signals in the H5'/5" region, these signals are not very useful. However, the H3' signals are typically sufficiently resolved to detect the decoupling effect. Observation of the H3' region upon irradiation of a ^{31}P signal should affect the signal of the H3' on the sugar 5' to that particular phosphate group. There are three drawbacks to this approach to ^{31}P signal assignments. First, unless one is dealing with a very small oligonucleotide, only the unusually shifted ^{31}P signals are sufficiently resolved for selective irradiation. Second, the H3' shift region contains the HDO signal and many H3' signals may be hidden underneath this signal. This problem may be overcome with difference spectroscopy by subtraction of proton spectra with and without ^{31}P irradiation. Third, in order to assign the ^{31}P resonances, the H3' resonances must be assigned. NOESY and/or COSY techniques can provide these assignments, but the H3' signals can be overlapped, leading to ambiguity. Despite these drawbacks, den Hartog et al. (65) were able to assign three ^{31}P resonances in **1**. The furthest downfield signal was assigned to the G5pG6 phosphate, the furthest upfield signal to C4pG5, and the T3pC4 ^{31}P signal falls within the normal range.

The effects of Pt drugs on shorter oligonucleotides containing adjacent G residues have been studied. The ^{31}P spectra of the cis-Pt(NH3)2Cl2 reaction products with d(CpGpG) (68), d(GpG) (69), and d(pGpGpG) (70) all exhibited a downfield shifted signal. For d(CpGpG)•cis-Pt(NH3)2, this signal occurs at ca. –2.6 ppm and is assigned via selective irradiation as the phosphate group between the two bound guanine residues. The ^{31}P signal of d(GpG)•cis-Pt(NH3)2 is shifted downfield but only to –3.3 ppm. We have attributed this small shift to the lack of a 5' phosphate group available for hydrogen bonding (71). Platination of d(pGpGpG) yields a major and a minor product. The major product was found to be [d(pGpGpG)•cis-Pt(NH3)2–N7(1),N7(2)] via pH-dependent chemical shift data in combination with comparison of the H1' splitting patterns. The three ^{31}P signals are assigned by analogy with previous assignments of single-stranded oligonucleotides.

In a recent ^{31}P study of the self-complementary oligonucleotide d(AGGCCT) and its adduct formed with cis-Pt(NH3)2Cl2 (72), platination was found to occur at the GG site and one downfield shifted ^{31}P signal at –2.9 ppm was reported. This signal was assigned as the phosphate group between the bound G residues by selective irradiation.

^{17}O Labeling. In this lab, a more thorough assignment of the ^{31}P signals of **1** was attempted using the ^{17}O-labeling method (Fig. 1). The oligonucleotide **1** was labeled at seven positions, allowed to react with cis-Pt(NH3)2Cl2, and studied by ^{31}P NMR. In this way, all four unusually shifted signals and three within the normal range were assigned. The two downfield ^{31}P signals at –2.60 and –3.10 ppm are assigned to the Pt binding site G5pG6 and to C2pT3, respectively. The upfield signal at –4.64 ppm is attributed to the T3pC4 moiety. T2pC3 and T7pC8 ^{31}P signals occur in the normal range. The assignment of the G5pG6 signal is consistent with that of den Hartog et al. (vide supra) (65). Since den Hartog et al. also assigned the upfield signal at –4.77 ppm, all four unusual signals have been assigned, as well as three within the normal range. Of some interest, only one unusually shifted signal is due to a direct binding effect of the Pt moiety. The remaining three signals are due to phosphate groups remote from and 5' to the binding site. This finding suggests some type of distorted secondary structure for the oligonucleotide induced by binding to the anti-tumor active Pt compound. A similar conclusion was reached earlier (65). However, the ^{17}O-labeling experiments allow us to suggest that the distortion occurs on the 5' end of the oligonucleotide. This may be due to hydrogen bonding interactions with the NH groups of the positive Pt moiety. It is noteworthy that only the C4pG5 and G5pG6 signals have unusual shifts at elevated temperature, suggesting disruption of the

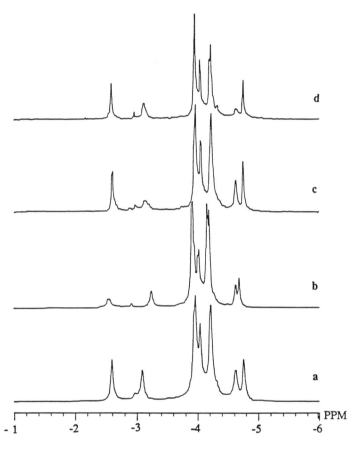

Figure 1. ^{31}P NMR spectra of ^{17}O-labeled and unlabeled $d(T_1C_2T_3C_4G_5$-$G_6T_7C_8T_9C_{10}) \cdot Pt(NH_3)_2$. a) no label, b) $G_5[^{17}O]G_6$, c) $C_2[^{17}O]T_3$, d) $T_3[^{17}O]C_4$.

conformation remote from the Pt binding site at high temperature, but retention of an unusual conformation at C_4pG_5 even at elevated temperatures.

HMQC. A new approach to the assignment of NMR resonances is the [1]H-detected heteronuclear multiple-quantum coherence (HMQC) two-dimensional correlation spectroscopy technique (32–39). The earliest reported example of this approach for ^{31}P NMR spectroscopy was a study conducted on the platinated oligonucleotide d(TGGT)•Pt(en) (40, 71). The HMQC experiment correlates each ^{31}P resonance with coupled ^{1}H resonances, namely H3' and H5'/5". These ^{1}H signals can be assigned *via* [1]H 2D-NOE methods. Due to the overlapping of signals prevalent in the H5'/5" region, connectivities with the H3' signals are most useful in ^{31}P NMR assignments. The ^{31}P NMR spectrum of this system consists of two signals at *ca.* –4.3 ppm and one downfield signal at –3.0 ppm. By correlation with the ^{31}P signal of the phosphate group 3' to each H3', the downfield ^{31}P signal was assigned to the Pt-bound GpG moiety and normal range signals to TpG and GpT. These assignments were confirmed via [17]O labeling. Similar methods were applied in the ^{31}P NMR assignments of d(TGGT)•*cis*-Pt(MeNH_2)_2 (71).

Summary. In all studies in which Pt unambiguously crosslinks two adjacent G's or in which Pt is suspected of crosslinking two adjacent G's or I's, there is a downfield shift of some ^{31}P NMR signal(s). The exact position of the shift depends on several factors, but for oligonucleotide systems with a NpGpG sequence, the signal is at *ca.* –3.0 ppm. In a recent comparison of the [31]P GpG shifts of several *cis*-PtA_2 adducts of d(TGGT), d(GGTT), and d(TTGG), we suggested (71) that hydrogen bonding between the 5' nucleotide and the amine is partly responsible for the downfield shifting, thus explaining the anomalously high field position of Pt crosslinked d(GpG) (69) and d(GpGTT) (71). Such H-bonding is revealed in X-ray structures (73, 74).

An Anomalous System: ^{31}P and ^{1}H NMR Investigation of d(TATGGGTACCATA)•Pt(en)

Previous Work. Recently in these laboratories, a combination of 2D ^{1}H and ^{31}P NMR techniques has been used to study the reaction of the deoxyribonucleotide d(TATGGGTACCATA), **2a**, with several drugs. Initial experiments with Pt drugs consisted mainly of 1D ^{31}P NMR experiments of the product obtained at low temperature where **2a** is largely in its duplexed form (75). A ^{31}P NMR spectrum of d(TATGGGTACCATA)•Pt(en), **2b**, [1.4 mM, pH 7.0, 25 °C, PIPES 10 buffer (0.01 M PIPES, 10^{-3} M EDTA, 0.1 M NaNO_3)] consists of seven closely spaced signals in the normal range of DNA (–4.0 to –4.5 ppm), two signals shifted considerably downfield (–2.57, –2.89 ppm), two signals slightly downfield (–3.67, –3.75 ppm), and two signals slightly upfield (–4.57, –4.70 ppm). The simplicity of this and the 1D ^{1}H NMR spectra, as well as electrophoretic analysis, indicate the presence of only one major product formed by the platination reaction (*75, 76*). This is an interesting result considering that there are two GpG, one GpNpG, one ApNpG, one GpNpA, and one ApNpA sites available for binding.

The ^{17}O-labeling technique was used to assign the ^{31}P signals of **2b**. The oligonucleotide is numbered in the following manner:

```
d(T  A  T  G  G  G  T  A  C  C   C   A   T  A)•Pt(en)
   1  2  3  4  5  6  7  8  9  10  11  12  13 14
```

The downfield signals at –2.57 and –2.89 ppm were assigned to the G$_5$pG$_6$ and the A$_8$pC$_9$ phosphate groups, respectively. The slightly downfield signals were assigned to G$_6$pG$_7$ and T$_3$pG$_4$, and the slightly upfield signals were assigned to A$_2$pT$_3$ and T$_7$pA$_8$, respectively.

The effects of temperature and pH on the ^{31}P spectrum of **2b** were studied (*75*). The signal at –2.89 ppm was found to be fairly insensitive to pH and temperature, whereas the other signals outside of the normal range were shifted into the main signal by high and low pH and high temperature. The observation of only one reaction product and more than one shifted ^{31}P signal indicate some type of unusual structure formed by platination that produces remote effects on phosphate groups other than that directly at the binding site. Due to its shift and its insensitivity to temperature and pH, it was suggested that the G$_5$pG$_6$ signal at –2.89 ppm indicates that Pt crosslinks G$_5$ and G$_6$. The shifting into normal range of the other ^{31}P signals, particularly the A$_8$pC$_9$ signal, was explained by the disruption of the unusual secondary structure by high temperature and non-neutral pH.

Other analytical methods were used in order to gain a better understanding of the nature of the structure of **2b** (*76*). Several lines of evidence, including CD spectroscopy, UV melting profiles, electrophoretic mobility, and ^{1}H NMR in H$_2$O, indicate that **2b** is not a duplex. The system has many properties similar to that of a hairpin, but there are enough differences that the term "hairpin-like" would be more appropriate.

Two important features of the aromatic region (7–9 ppm) of the ^{1}H NMR spectrum should be noted. First, there are two broad signals that are shifted downfield. This type of signal is typical of the H8 of platinated guanine bases in single-stranded oligonucleotides (*66, 68, 70, 71, 77, 78*). A strong NOE crosspeak between these two signals is another feature characteristic of H8 signals of crosslinked species. Second, 18 signals are expected in this region. However, only 17 are observed. T$_1$ measurements allow the assignment of the four A H2 signals due to their long T$_1$ values (*79, 80*). The three C H6 doublet signals are readily identified. T H6 signals are typically the furthest upfield in the aromatic region (*79, 80*), and so the four upfield singlets are most probably T H6's. The remaining six signals are assigned as H8's due to their ability to exchange with D$_2$O (*81*). Since there should be seven H8 signals present, the missing signal was assigned as an H8 (75).

The results at that time suggested a hairpin-like structure in which the 5' end TATG hydrogen bonds with the 3' end CATA and the loop containing the sequence GGTACC is stabilized by hydrogen bonds between the amine groups on the Pt and the oligonucleotide. Since the Pt was probably bound at the G_5pG_6 site, the two broad downfield ^1H signals were tentatively assigned as G_5 and G_6 H8's. The missing signal was unexplained. This initial hairpin-like model is called Model I here.

Further Studies. The collection of properties that led to Model I is certainly unusual. The uniqueness of the results was intriguing, and we have continued to investigate this system. Conceivably yet unrecognized structural or Pt anticancer drug chemistry would be revealed if more details of the nature of **2b** could be understood. Furthermore, the remarkable high yield formation of one product is very surprising, given the alternative binding sites enumerated above. Additionally, the solution of this structure represents a significant challenge to modern NMR methods.

With ^1H NMR spectra recorded at 500 and 600 MHz, we have confirmed the absence of one aromatic signal as deduced earlier at 360 MHz. Therefore, the missing signal is probably not due to signal overlap. Model I assumes an intact oligonucleotide. However, one possible explanation for the missing aromatic signal would be depurination caused by the platination reaction. In order to test this possibility, a sample of **2b** was treated with KCN to remove the platinum moiety from the oligonucleotide. The oligonucleotide was purified by HPLC methods. Digest analysis of the stripped oligomer produced a ratio of 4 A: 4 T: 3 G: 3 C, thereby eliminating the possibility of depurination (82). Furthermore, one consequence of depurination is to produce a mixture of a and b anomers of 2'-deoxyribose, which exist in a tautomeric equilibrium with the ring-opened aldehyde (83). All H1' signals are normal, also ruling out depurination.

In order to assign signals in the ^1H spectrum, 2D NMR techniques were employed on the $Pt(en)Cl_2$ reaction product of a slightly shorter version of **2a**, i.e., d(ATGGGTACCCAT) **3a**. The resultant platinated oligonucleotide, d(ATGGGTACCCAT)•Pt(en), **3b**, has ^{31}P NMR, ^1H NMR (H_2O and D_2O), UV temperature dependence, and electrophoresis characteristics similar to that of **2b**. Of course, there are fewer ^1H and ^{31}P NMR signals. Of particular note, one aromatic signal is missing.

To remain consistent with the numbering scheme of the tetradecadeoxyribonucleotides, **2a** and **2b**, we number **3b** as follows:

d(A T G G G T A C C C A T) •Pt(en)
   2 3 4 5 6 7 8 9 10 11 12 13

Initially, 2D NOE and COSY experiments at 500 MHz confirmed our earlier conclusion, i.e., that the Pt adducts have unusual structural features and are not duplex B-form DNA. No evidence was observed for another aromatic signal, and indeed, many connectivities of duplex DNA molecules were missing. However, by a combination of NOESY and COSY, some non-specific assignments were made or confirmed. NOESY and COSY connectivities are seen between the T methyl groups and the three most upfield signals in the aromatic region, thereby confirming their assignments as T H6's as the most upfield aromatic singlets. These results further confirm the identity of the missing signal as that of an H8.

Since many crosspeaks are missing in comparison to B-form DNA, we chose to gain an entry into specific ^{1}H NMR assignments by several routes. Points of entry include: (a) HMQC and SRCSC ^{31}P–^{1}H correlation spectroscopies; (b) single frequency ^{31}P decoupling; and (c) the assumption that the ends were duplexed in a right-handed helix allowing limited use of standard internucleotide NOE's. Several of these approaches were combined to present a consistent picture of the assignments.

The more specific ^{1}H NMR assignments were initiated by use of ^{17}O-labeling results combined with the HMQC ^{31}P–^{1}H correlation experiment (40) (Fig. 2). For example, the furthest downfield ^{31}P signal of **3b** is assigned to A_8pC_9 by analogy to the ^{17}O-labeling experiments with **2b**. This signal is coupled to an H3' at 5.14 ppm which has a moderately strong NOE crosspeak to the broad downfield signal at 8.65 ppm. This H3' also has COSY connectivities through H2'/2" protons to an H1' signal at 6.05 ppm that has an NOE connectivity to the same downfield 8.65 ppm signal. On the basis of these results, the signal at 8.65 ppm is assigned as A_8 H8. Using this method, the T_3 and T_7 H6 signals were also assigned. The remaining T H6 could then be assigned to T_{13}.

In this way, we were able to assign three of the six H3' signals resolved between 4.65 and 5.15 ppm. The most downfield H3' signal at 5.12 ppm has an NOE crosspeak to an H8 signal tentatively assigned to G4 (see below). The signal at 5.0 ppm has an NOE crosspeak with an H8 signal at 8.26 ppm. This signal has an NOE crosspeak to a T methyl signal that is assigned to T_{13}. Therefore, the H3' and H8 signals are assigned to A_{12}. Similarly, the remaining resolved signal clearly is for H3' of A_2 because of an NOE crosspeak to A_2 H8 assigned in the same manner as A_{12} H8. The remaining five H3' signals in the region are for the three C's and G_5 and G_6. Finally, the H3' for T_{13} is observed at 4.54 ppm by its NOE crosspeak to T_{13} H6.

The region containing the H5'/5" signals is crowded, so assignments using the ^{31}P–H5'/5" connectivities are difficult. Although HMQC provided useful information, a limitation encountered in this experiment is that no scalar connectivities were observed with the G_5pG_6 and G_4pG_5 ^{31}P signals at 21 °C. The G_6pT_7 signal has a clear connectivity, but to the severely overlapped H3' region at ~4.8 ppm.

Given the observation that the overlapping signal at 4.8 ppm contains the three C and two G H3' signals, the following arguments assign the H3' signals to G_5 and G_6. The HMQC experiment identifies one of these H3'

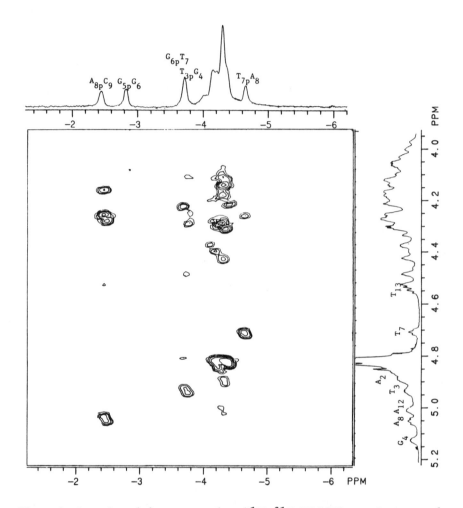

Figure 2. A section of the contour plot of ^1H–^{31}P HMQC experiment on the platinated oligonucleotide $d(A_2T_3G_4G_5G_6T_7A_8C_9C_{10}C_{11}A_{12}T_{13})\bullet Pt(en)$ (0.002 M, 99.9% D_2O, 10^{-4} M EDTA, pH 7, 21 °C). Pulse sequence used was $90(^1H)$– $1/2J_{HP}$–$90(^{31}P)$–$t_1/2$–$180(^1H)$–$t_1/2$–$90(^{31}P)$–acquire(^1H) (36). 1/2 J delay was set at 50 ms. ^1H sweep width was 3012 Hz in 512 points for an acquisition time of 0.26 s. A repetition delay of 1.3 s was used. 128 blocks of 128 scans were collected. The sweep width in the ^{31}P dimension was 1706 Hz with an effective acquisition time of 42.6 ms. The ^{31}P dimension was zero-filled once before processing. Data are displayed in absolute value mode. Data were collected on a GN-500 spectrometer equipped for reverse detection experiments.

to be that for G_6. A selective irradiation (decoupling of the ^{31}P signal at –2.8 ppm assigned to P5 by ^{17}O labeling) demonstrates that the G_5 H3' is also in this region. Therefore, the H3' at 5.13 ppm is assigned to G_4, as mentioned above. This H3' signal has an NOE crosspeak to an H8 signal at 8.18 ppm. This H8 signal also has crosspeaks to the H1', H3' and H2' or H2" on the deoxyribose of T_3, substantiating these assignments and the conclusion that G_4 does not bind Pt.

Assuming the aromatic proton signal assignments are correct thus far, it remains that the most downfield H8 signal is for G_5 or G_6. This base is in an unusual conformation since the H8 has only one NOE, and that is a strong one to its H1'. Since this suggests that the G is in a *syn* conformation (2), we refer to it as G_S. Of the remaining eleven H1' signals, six have NOE crosspeaks to an AH8 or a TH6. Two have NOE crosspeaks to a CH6. These two and a third CH6 signal have NOE crosspeaks to H3' signals, which have COSY connectivities to H1' signals. Thus, the 9 H1' signals for the three A's, three T's and three C's are identified. The G_4 H8 has a crosspeak to an H1' in addition to the H3' signal discussed above. The G_4 H1' and H3' signals are connected by COSY connectivities. Therefore, the H1' that has no crosspeaks to an aromatic signal must be for G_5 or G_6.

This H1' of the G base with the missing H8 (G_m) has a crosspeak to a signal in the H4'/5'/5" region (most probably the H4' of its own sugar). This signal has a weak crosspeak to the T_7 methyl signal. The H6 and H3' signals of this T_7 have crosspeaks to the H1' of the G with the *syn* conformation. In turn, the H2'/2" signal of this deoxyribose has a weak crosspeak to the T_7 methyl. Therefore, there are crosspeaks between T_7 and signals of *both* unusual G moieties, the one with the *syn* conformation and the one with the missing H8 signal. Although the *syn* G has a strong NOE crosspeak to A_8H8, this information is not enough to distinguish G_5 from G_6 signals.

Although the additional two nucleotide units in **2b** compared to **3b** complicate the ^1H NMR assignments there is slightly less overlap in the ^{31}P spectrum of the 14-mer, possibly because of lower conformational flexibility of the larger oligomer due to the longer stem of the hairpin-like molecule. We decided to exploit this feature of the ^{31}P spectrum of d(TATGGGTACCCATA)•Pt(en) by using the selective ^{31}P–^1H chemical shift correlation (SRCSC) experiment (29). Also, due to the complete assignment of the ^{31}P NMR signals for **2b** and the resolution of the crosspeaks in the SRCSC experiment (as compared to the HMQC experiment), more H3' signals both within the overlapping region at *ca.* 4.83 ppm and outside this region may be assigned independently of NOE results. In particular, the G_4 H3' signal can be unambiguously assigned. The SRCSC experiment on **2b** was conducted at 12 °C, the temperature of the NOESY experiment, and where more (twelve) ^{31}P–^1H crosspeaks could be resolved (Fig. 3A).

At 12 °C, the furthest downfield ^{31}P NMR signal at –2.6 ppm (seen folded over at –4.40 ppm in Fig. 3A), assigned as A_8pT_7 by ^{17}O-labeling, is coupled to an H3' at 5.05 ppm. This H3' has an NOE crosspeak to the broad downfield ^{1}H signal at 8.68 ppm (12 °C), thus assigning these two signals (H3' and H8) as A_8. This method was used to assign the H3' signals of A_2 and T_7 at 5.08 and 4.73 ppm as well as their respective H8 and H6 signals (8.52 and 7.22 ppm) (12 °C). The two ^{31}P signals at –3.67 and –3.75 ppm (0.1 M NaNO$_3$, 25 °C) were assigned to G_6pT_7 and T_3pG_4, respectively (75,76). At 12 °C (no salt), these two signals are overlapped and only one ^{31}P–^{1}H crosspeak is observed.

At 40 °C (no salt), signals are observed at –3.78 and –3.85 ppm. The –3.78 ppm signal exhibits coupling to an H3' at 4.93 ppm (40 °C) which has an NOE crosspeak to an unassigned T H6 at 7.34 ppm (12 °C). On this basis, these two proton signals (H3' and H6) are assigned as T_3. The –3.85 ppm signal, which must then be G_6pT_7, exhibits a crosspeak to an H3' at 4.80 ppm (within the overlapping region). At 12 °C, G_6 H3'–^{31}P coupling is either not observed or the T_3 and G_6 H3' signals are coincident. These results indicate that differing salt concentration leads to reversal in shift order of the two ^{31}P signals.

As for **3b**, a NOESY crosspeak was observed between the A_2 H8 and T_3 Me signals in **2b**. A similar crosspeak was observed between unassigned H8 and T Me signals, thus identifying them as A_{12} H8 and T_{13} Me. A COSY crosspeak to T_{13} Me assigns T_{13} H6. With the T_3, T_7, and T_{13} H6 signals assigned, the unassigned T H6 at 7.38 ppm must be T_1. T_1 H6 has a NOE crosspeak to a H3' signal at 4.70 ppm (12 °C) which is coupled to a ^{31}P signal in the normal region.

A heteronuclear 2D J experiment (30) which yields the H3'–C–O–P coupling constants was conducted on **2b** (Fig. 3B). It was necessary to conduct this experiment at 40 °C due to interference of the HDO signal at lower temperatures. Two H3' signals exhibiting no observable coupling to ^{31}P were observed at 4.71 and 4.84 ppm (40 °C). A_{14} has no 3' phosphate group since it is at the 3' end. The H3' signal at 4.71 ppm is assigned to A_{14} for two reasons. First, the H3' signal of the most 3' base is generally found to occur upfield of most or all other H3' signals (13, 72). Second, this H3' has an NOE crosspeak to an H8 at 8.24 ppm (12 °C) assigned to A_{14} on the basis of an NOE to T_{13} H1' (see below). The G_5pG_6 ^{31}P signal at –3.12 ppm (40 °C) exhibits <1 Hz coupling to an H3'. However, by analogy to **3b**, the uncoupled H3' in the overlapping region at ca. 4.83 ppm (40 °C) is assigned as G_5. The G_4 H8 at 8.15 ppm was assigned on the basis of an NOE to T_3 H1' (see below). An NOE crosspeak to the H3' at 5.12 ppm assigns this signal to G_4. The C H6 signal at 7.58 ppm has an NOE to the isolated H3' at 4.87 ppm (12 °C). This H3' is coupled to a normal region ^{31}P signal at –4.34 ppm (40 °C) which ^{17}O-labeling indicates is $C_{11}pA_{12}$. These two signals (H6 and H3') are therefore assigned to C_{11}. ^{31}P NMR signals assigned by ^{17}O labeling as C_9pC_{10} (–4.23 ppm, 40 °C) and $C_{10}pC_{11}$ (–4.83 ppm, 40 °C) have ^{31}P–^{1}H crosspeaks (12 °C and 40 °C) to two overlapping H3' signals at 4.83 ppm. One C H6 (C_{10}, see below) has an NOE to a signal in this region. Thus, C_9 H3' and C_{10} H3' are at ca. 4.83 ppm. The

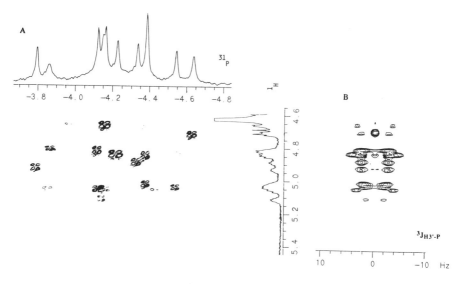

Figure 3. Sections of the contour plots of A) ^1H detected ^1H–^{31}P selective reverse chemical shift correlation experiment; B) ^1H detected ^1H–^{31}P heteronuclear 2D J correlation map of $d(T_1A_2T_3G_4G_5G_6T_7A_8C_9C_{10}C_{11}A_{12}T_{13}A_{14})•Pt(en)$ (0.003 M, 99.9% D_2O, 10^{-4} M EDTA, pH 7, 40 °C). Normal 1D ^1H and ^{31}P spectra are shown along respective axes for reference. Experimental conditions in A were: saturation(^1H)–90(^{31}P)–$t_1/2$–180_{sel}(^1H)–180(^1H)–$t_1/2$–90(^{31}P)–90(^1H)– acquire(^1H) (30). Sweep width in the ^1H dimension was 1300 Hz in 1024 points for an acquisition time of 0.787 s. 128 blocks of data of 128 scans were collected. The magnitude of the selective pulse was 125 Hz, with the transmitter set at 4.8 ppm. Alternate scans were stored in separate areas of memory and processed to yield a phased 2D spectrum (30). The sweep width in the ^{31}P dimension was 300 Hz with an effective acquisition time in that dimension of 426 ms. The ^{31}P dimension was zero-filled twice before processing. The resulting phased 2D spectrum including both positive and negative contours is shown. This experiment selects for H3'–P correlations. The splitting in the ^{31}P dimension is the vicinal coupling between these nuclei. Spectral range in the ^{31}P dimension was reduced to increase digital resolution resulting in the folding over of one peak at –4.25 ppm ^{31}P and 5.05 ppm ^1H. Experimental conditions for B were 90(^1H)–$t_1/2$– 90_{sel}(^1H)–180(^{31}P)–90_{sel}(^1H)–$t_1/2$–acquire(^1H)decouple(^{31}P) (30). The ^1H sweep width was 1300 Hz with a size of 1024 points for an acquisition time of 787 ms. 64 blocks of 64 scans each were collected. The magnitude of the selective pulse was 119 Hz centered at 4.75 ppm. The spectral width in the f_1 dimension was 50 Hz with an effective acquisition time of 1.28 s. The f_1 dimension was zero-filled twice before processing. Data are presented in phased mode. All data in A and B were obtained with a GN-500 spectrometer equipped for reverse detection experiments.

H3' assignments made for **2b** by use of the SRCSC, 2D J, and NOESY experiments are fully consistent with those of **3b** (*vide supra*). As for **3b**, the assignments thus far identify the furthest downfield broad H8 signal as G_5 or G_6. A strong NOE is observed between this signal and A_8 H8.

Table I. H3' H-P Coupling Constants (Hz) and Chemical Shifts (ppm)
for d(TATGGGTACCCATA)•Pt(en)

H3'	Coupling Constants[a]	Chemical Shift[b] 12 °C	40 °C
T_1	6.6	4.70	4.66
A_2	3.3	5.08	5.03
T_3	5.0	4.95	4.91
G_4	3.4	5.12	5.12
G_5	<1	4.85[c]	4.84
G_6	4.7	4.95[d]	4.80
T_7	5.2	4.73[e]	4.71
A_8	6.9	5.05	5.05
C_9	8.2	4.83	4.83
C_{10}	4.9	4.82	4.84
C_{11}	5.1	4.87	4.88
A_{12}	3.7	5.03	5.02
T_{13}	5.3	4.84	4.81
A_{14}	<1	4.74	4.71

[a] Values obtained from 2D J coupling experiment at 40 °C. Estimated accuracy of ±0.15 Hz.

[b] Values obtained from $^{31}P-^{1}H$ SRCSC experiment except where indicated.

[c] Approximate value.

[d] Tentative value only; G_6 H3'–^{31}P coupling is either not observed or G_6 H3' is coincident with T_3 H3' (value shown).

[e] $^{31}P-^{1}H$ coupling is difficult to observe at 12 °C. Chemical shift value determined from NOESY experiment.

The heteronuclear 2D J coupling experiment permits measurement of $^{3}J_{HCOP}$ couplings between ^{31}P and H3' nuclei (Fig. 3B). Table I contains the $^{3}J_{HCOP}$ coupling constants measured for each base H3'. Many of the coupling constants listed are significantly greater or less than those observed for typical duplexed DNA and oligodeoxyribonucleotides (3–5 Hz) (*16, 30*) indicating distortions in the phosphate backbone. A Karplus relationship correlating $^3J_{HCOP}$ with the H-C-O-P dihedral angle has been proposed by Lankhorst et al. (*84*):

$$J_{HCOP} = 15.3 \cos^2 \phi - 6.1 \cos \phi + 1.6$$

This equation is a multi-valued function, resulting in four possible ϕ values for any given J. Since the platinated oligonucleotides **2b** and **3b** have unusual structures and we do not, at this point, have a definite model of these structures, none of these calculated ϕ values can be discarded. However, the ϕ values calculated for each phosphate group are valuable as constraints in any future molecular mechanics calculations, especially since there is a deficiency of internucleotide NOE's in this system. It is of interest that the most atypical values are those for G_5 (<1 Hz), A_8 (6.9 Hz), and C_9 (8.2 Hz), all bases within the loop of the proposed hairpin-like structure.

Several possible modes of platination are being considered to evaluate the data we have collected on **2b** and **3b**. In our original Model I (Fig. 4), described above, a normal G_5G_6 crosslink would explain the insensitivity of the G_5pG_6 ^{31}P signal to pH and temperature. The other unusual ^{31}P signals would be due to the hairpin-like structure and thus would be sensitive to temperature and pH. The missing signal could be a consequence of broadening of an H8 due to a conformational process. However, this model does not explain the strong NOE crosspeak between the two broad downfield H8 signals, $G_{(5 \text{ or } 6)}$ and A_8. Strong NOE's are usually observed between H8 signals of platinated purine bases (*1, 71, 77*). Also, the A_8 H8 signal has characteristics typical of a platinated purine (*66, 68, 70, 71, 77*). There are no reasons for believing that AH8 would be shifted downfield if it were not coordinated to Pt. Thus, our original model, based on limited information, appears to be partly incorrect unless some unusual rearrangement places the sugar of A_8 near the H8 of G_5 or G_6. In such a case, our assignment of the 8.65 ppm signal to A_8 H8 is in error, but we believe that such an error is unlikely.

In a second type of model, Model II, Pt is bound to three bases, G_5, G_6, and A_8 (Fig. 5). This would require displacement of one binding site of the en group or the formation of a five-coordinate complex. The pH and temperature insensitivity of the ^{31}P NMR signal of the G_5pG_6 is still explained by a direct effect of Pt crosslinking G_5 and G_6. The A_8pC_9 group is outside the Pt binding site and so would be sensitive to secondary structure disrupting effects. The A_8 and one platinated G H8 would be brought close enough to account for the NOE between the two downfield signals. The other platinated G H8 signals could be missing for two possible reasons. First, the H8 is broadened by Pt binding at N7. Second, the free displaced end of the en group is situated near the H8 proton and catalyzes its exchange with H_2O.

In a variation of Model II, one of the G bases actually forms a Pt–C8 bond. This model, Model IIC (for carbon), could explain some of our data.

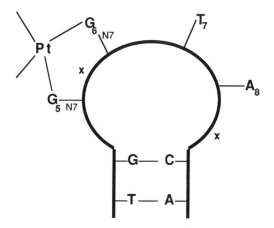

Figure 4. Model I proposed for the structure of the TGGGTACCCA region of **2b** *and* **3b**. *X indicates a phosphate group with an unusually shifted* [31]*P signal.*

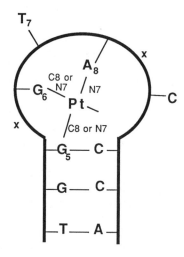

Figure 5. Model II proposed for the structure of the TGGGTACCCA region of **2b** *and* **3b**. *X indicates a phosphate group with an unusually shifted* [31]*P signal.*

It would explain the absence of one GH8 signal, whereas we find the H3', H1', and ^{31}P signals for all nucleotides. The problem with this model is that the en would have to dissociate at one end – a problem with Model II. The Pt–C bond would have to be labile to CN⁻ cleavage.

Two sets of results provide evidence against Model II. First, the ^{195}Pt NMR spectra of **2b** and **3b** have signals at –2645 and –2640 ppm, respectively. These chemical shifts are similar to the reported shift for [Pt(en){(Guo)$_2$-N7,N7}]$^{2+}$ (–2662 ppm) (*85*). This similarity strongly suggests that Pt is bound to two purine bases via N7. Binding to a C (such as C8 in Model IIC) would shift the ^{195}Pt NMR signal *ca.* 300 ppm upfield (*86*). Second, upon addition of KCN and the resulting removal of Pt from **3b**, the spectrum of the unreacted duplex was regenerated. All 15 signals were present, indicating that the proton of the missing signal in Pt(en)•[d(ATGGGTACCCAT)], or any moiety containing that proton, had not been physically removed from the oligonucleotide by platination. The presence of all signals in the D$_2$O solution (Fig. 6) also indicates that Pt does not bind at any C8 position.

We propose an alternative two-site binding model with A$_8$ and G$_5$ or G$_6$ bound to Pt, Model III (Fig. 6). Either possibility explains the downfield G$_5$pG$_6$ signal. If G$_s$ is G$_6$, a downfield ^{31}P signal is expected since, in Z-DNA, the G residues exist in a *syn* conformation and the ^{31}P signals of the phosphate groups immediately 5' to these residues are shifted downfield (*16*). If G$_s$ is G$_5$, the unusual position of the G$_5$pG$_6$ ^{31}P signal would be explained as a direct effect of Pt binding due to its location within the region bound by Pt.

G$_5$ binding is indicated by the lack of an upfield shift for T$_7$ H6 of **3b** between pH 8 and 10. Other results also suggest binding at G$_5$. First, the pH dependence study revealed an upfield shift for C$_{10}$ H6 from pH 7.9 to 10.2. Assuming that G$_s$ is G$_5$, this result could be explained by deprotonation of N1 H of G$_5$ (downfield shift of G$_s$ H8 at *ca.* pH 8) and disruption of the G$_5$C$_{10}$ base pair. Second, since G$_s$ is bound to Pt, we expect a decreased pK_a to *ca.* 8. A pK_a of *ca.* 9.5 is actually observed. However, if the G is hydrogen-bonded to C$_{10}$, as G$_5$ is, the pK_a will be higher.

A possible explanation for the missing H8 signal is that the Pt moiety is close to the H8 proton, resulting in a large change in chemical shift for that proton signal. Vibrational effects would result in severe broadening of this ^{1}H signal so that it is not observable. The presence of some unusual NOE connectivities indicates that the loop may fold over into a region near the stem. In particular, NOE's were observed between the imino signal of G$_4$C$_{11}$ (within the base stem) and signals of C$_9$ H6, G$_s$ H8, and G$_m$ H1' in 1D NOE studies. NOE's were also observed between A$_8$ H8 and G$_5$C$_{10}$ and between C$_{11}$ NH$_a$ and A$_8$ H2. The bend is probably more severe than that reported for platinated duplexes (*77, 87, 88*).

In summary, we have presented three possible models to explain the results we have obtained on this system. We feel, however, that Model III, in which Pt is bound to G$_5$ and A$_8$, is the most reasonable (Fig. 7). The

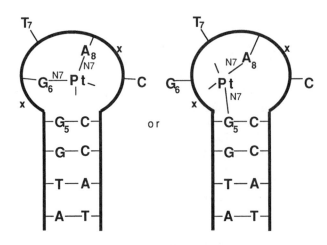

Figure 6. Model III proposed for the structure of Pt(en)•[d(ATGGGTACCCAT)] and Pt(en)•[d(TATGGGTACCCATA)]. X indicates a phosphate group with an unusually shifted ^{31}P signal.

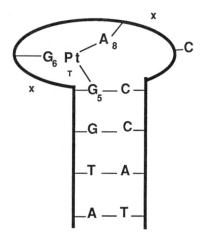

Figure 7. Final model proposed for structure of Pt(en)•[d(ATGGGTACCCAT)] and Pt(en)•[d(TATGGGTACCCATA)] showing approximate locations of nucleotides.

regions T_1 through G_5 and C_{10} through A_{14} base pair with each other to form the stem region. The T_7 base moiety is located within the loop region and is in close proximity to both G_5 and G_6 nucleotides. The C_9 base moiety is forced out of the loop. G_5 exists in a *syn* conformation. The loop region may bend over, bringing some protons in this region into close proximity to some stem region protons.

Future Studies

It is clear that, regardless of the eventual outcome of these studies, an unprecedented structure will emerge. Significant progress has been made since we last reported on this system (76), but additional studies are called for, based on our recent 2D NMR results. Clearly, the issue of the presence or absence of the bound en must be addressed. We have not found signals unambiguously attributable to this ligand. Perhaps a $MeNH_2$ complex will answer this question. The CN^- displacement reaction, if carried out in D_2O, should place a D at position 8 of the affected base if a Pt–C bond is present and broken during CN^- displacement. The replacement of C with 5-methylC could help to assign this evidently "floppy" region. A severe limitation continues to be spectral overlap of the H3' signals, which has prevented a specific assignment of which G is bound by Pt, i.e. G_5 or G_6. Eventually, this overlap may be resolved by temperature variation studies.

Acknowledgments

This research was supported by NIH Grant GM 29222 to L. G. M. The NMR instrumentation was purchased with funds awarded by NIH and NSF. We are grateful to these organizations. We wish to thank Dr. Robert Jones (Emory University, Atlanta, GA) for his assistance with the GN-500 spectrometer and Scott Davidson (Applied Biosystems, Foster City, CA) for his digest analysis work.

Literature Cited

1. Reedijk, J.; Fichtinger-Schepman, A. M. J.; van Oosteroom, A. T.; van de Putte, P. *Structure and Bonding* **1987**, *67*, 53.
2. Wüthrich, K., *NMR of Proteins and Nucleic Acids*; John Wiley and Sons, Inc.: New York, **1986**; and references contained therein.
3. Reid, B. R. *Quart. Rev. Biophysics* **1987**, *20*, 1.
4. Patel, D. J.; Shapiro, L.; Hare, D. *Quart. Rev. Biophysics* **1987**, *20*, 35.
5. Feigon, J.; Wright, J. M.; Leupin, W.; Denny, W. A.; Kearns, D. R. *J. Am. Chem. Soc.* **1982**, *104*, 5540.

6. Feigon, J.; Leupin, W.; Denny, W. A.; Kearns, D. R. *Biochemistry* **1983**, *22*, 5943.

7. Haasnoot, C. A. G.; Westerink, H. P.; van der Marel, G. A.; van Boom, J. H. *J. Biomol. Struct. Dyn.* **1983**, *1*, 31.

8. Hare, D. R.; Wemmer, D. E.; Chou, S. H.; Drobny, G.; Reid, B. R. *J. Mol.. Biol.* **1983**, *171*, 319.

9. Jamin, N.; James, T. L.; Zon, G. *Eur. J. Biochem.* **1985**, *152*, 157.

10. Suzuki, E.; Pattabiraman, N.; Zon, G.; James, T. L. *Biochemistry* **1986**, *25*, 6854.

11. Chary, K. V. R.; Hosur, R. V.; Govil, G.; Zu-kun, T.; Miles, H. T. *Biochemistry* **1987**, *26*, 1315.

12. Hare, D. R.; Reid, B..R. *Biochemistry* **1986**, *25*, 5341.

13. Scott, E. V; Zon, G.; Marzilli, L. G.; Wilson, W. D. *Biochemistry*, **1988**, *27*, 7940.

14. Scott, E. V.; Jones, R. L.; Banville, D. L.; Zon, G.; Marzilli, L. G.; Wilson, W. D. *Biochemistry* **1988**, *27*, 915.

15. Jones, R. L.; Scott, E. V.; Zon, G.; Marzilli, L. G.; Wilson, W. D. *Biochemistry*, **1988**, *27*, 6021.

16. Gorenstein, D. G., Ed., *Phosphorus-31 NMR*; Academic Press, Inc.: New York, **1984**; and references contained therein.

17. Lankhorst, P. P.; Erkelens, C.; Haasnoot, C. A. G.; Altona, C. *Nucleic Acids Res.* **1983**, *11*, 7215.

18. Borer, P. N.; Zanatta, N.; Hodak, T. A.; Levy, G. C.; van Boom, J. H.; Wang, A. H. -J. *J. Biomol. Struct. Dyn.* **1984**, *2*, 1373.

19. Leupin, W.; Wagner, G.; Denny, W. A.; Wuthrich, K. *Nucleic Acids Res.* **1987**, *15*, 267.

20. Sklenar, V.; Bax, A.; Zon, G. *J. Am. Chem. Soc.* **1987**, *109*, 2221.

21. Ashcroft, J.; Live, D. H.; Patel, D. J.; Cowburn, D. *Biopolymers*, submitted for publication.

22. Gao, X.; Jones, R. A. *J. Am. Chem. Soc.* **1987**, *109*, 1275.

23. Gao, X.; Jones, R. A. *J. Am. Chem. Soc.* **1987**, *109*, 3169.

24. Saenger, W. *Principles of Nucleic Acid Structure*; Springer-Verlag: New York, **1984**.

25. James, T. L., In *Phosphorus-31 NMR*; Gorenstein, D. G., Ed. Academic Press, Inc.: New York, **1984**; Ch. 12.

26. Chen, C.- W.; Cohen, J. S., In *Phosphorus-31 NMR*; Gorenstein, D. G., Ed. Academic Press, Inc.: New York, **1984**; Ch. 8.

27. Cheng, D. M.; Kan, L.- S.; Miller, P. S.; Leuzinger, E. E.; Ts'o, P. O. *Biopolymers* **1982**, *21*, 697.

28. Frey, M. H.; Leupin, W.; Sorensen, O. W.; Denny, W. A.; Ernst, R. R. *Biopolymers* **1985**, *24*, 2381.

29. Sklenar, V.; Miyashiro, H.; Zon, G.; Miles, H. T.; Bax, A. *FEBS Lett.* **1984**, *208*, 94. (There is a misprint in the phase cyling for the experiment in this reference, refer to reference 31.)

30. Sklenar, V.; Bax, A. *J. Am. Chem. Soc.* **1987**, *109*, 7525.

31. Gorenstein, D. G.; Lai, K.; Shah, D. O. *Biochemistry* **1984**, *23*, 6717.

32. Muller, L. *J. Am. Chem. Soc.* **1979**, *101*, 4481.

33. Bodenhausen, G.; Reuben, D. *J. Chem. Phys. Lett.* **1980**, *69*, 185.

34. Bendall, M. R.; Pegg, D. T.; Doddrell, D. M. *J. Magn. Reson.* **1983**, *52*, 81.

35. Redfield, A. G. *Chem. Phys. Lett.* **1983**, *96*, 539.

36. Bax, A.; Griffey, R. H.; Hawkins, B. L. *J. Mag. Reson.* **1983**, *55*, 301.

37. Live, D. H.; Davis, D. G.; Agosta, W. C.; Cowburn, D. *J. Am. Chem. Soc.* **1984**, *106*, 6104.

38. Otvos, J. D.; Engeseth, H. R.; Wehrli, S. *J. Mag. Reson.* **1985**, *61*, 579.

39. Live, D. H.; Armitage, I. M.; Dalgarno, D. C.; Cowburn, D. *J. Am. Chem. Soc.* **1985**, *107*, 1775.

40. Byrd, R. A.; Summers, M. F.; Zon, G.; Fouts, C. S.; Marzilli, L. G. *J. Am. Chem. Soc.* **1986**, *108*, 504.

41. Griffey, R. H.; Redfield, A. G. *Quart. Rev. Biophysics* **1987**, *19*, 51.

42. Joseph, A. P.; Bolton, P. H. *J. Am. Chem. Soc.* **1984**, *106*, 437.

43. Petersheim, M.; Mehdi, S.; Gerlt, J. A. *J. Am. Chem. Soc.* **1984**, *106*, 439.

44. Connolly, B. A.; Eckstein, F. *Biochemistry* **1984**, *23*, 5523.

45. Ott, J.; Eckstein, F. *Biochemistry* **1985**, *24*, 2530.

46. Connolly, B. A.; Potter, B. V. L.; Eckstein, F.; Pingoud, A.; Grotjahn, L. *Biochemistry* **1984**, *23*, 3443.

47. Cohn, M.; Hu, A. *Proc. Natl. Acad. Sci. USA* **1978**, *75*, 200.

48. Lowe, G.; Sproat, B. S. *J. Chem. Soc. Chem. Commun.* **1978**, 565.

49. Lutz, D.; Nolle, A.; Staschewski, D. *Z. Naturforsch* **1978**, *A33*, 380.

50. Wilson, W. D.; Jones, R. L.; Zon, G.; Scott, E. V.; Banville, D. L.; Marzilli, L. G. *J. Am. Chem. Soc.* **1986**, *108*, 7113.

51. Marzilli, L. G.; Banville, D. L.; Zon, G.; Wilson, W. D. *J. Am. Chem. Soc.* **1986**, *108*, 4118.

52. Kopka, M. L.; Yoon, C.; Goodsell, D.; Pjura, P.; Dickerson, R. E. *Proc. Natl. Acad. Sci. USA* **1985**, *82*, 1376.

53. Patel, D. J.; Shapiro, L. *J. Biol. Chem.* **1986**, *261*, 1230.

54. Patel, D. J.; Canuel, L. L. *Proc. Natl. Acad. Sci. USA* **1979**, *76*, 24.

55. Davanloo, P.; Crothers, D. M. *Biopolymers* **1979**, *18*, 2213.

56. Keniry, M. A.; Brown, S. C.; Berman, E.; Shafer, R. H. *Biochemistry* **1987**, *26*, 1058.

57. Roberts, J. J., In *Advances in Inorganic Chemistry*; Marzilli, L. G., Eichhorn, G. L., Eds., Elsevier North Holland, Inc.: New York, **1981**; Vol. 3, p. 273.

58. Roberts, J. J.; Pera, M. F. *Am. Chem. Soc. Symp. Ser.* **1983**, *209*, 3.

59. Martin, R. B. *Acc. Chem. Res.* **1985**, *18*, 32.

60. Martin, R. B.; Miriam, Y. H. In *Metal Ions in Biological Systems*; Sigel, H., Ed., Marcel Dekker, Inc.: New York, **1979**; p. 57.

61. Fichtinger-Schepman, A. M. J.; Lohman, P. H. M.; Reedijk, J. *Nucleic Acids Res.* **1982**, *10*, 5345.

62. Fichtinger-Schepman, A. M. J.; van der Veer, J. L.; den Hartog, J. H. J.; Lohman, P. H. M.; Reedijk, J. *Biochemistry* **1985**, *24*, 707.

63. Wilson, W. D.; Heyl, B. L.; Reddy, R.; Marzilli, L. G. *Inorg. Chem.* **1982**, *21*, 2527.

64. Marzilli, L. G.; Reily, M. D.; Heyl, B. L.; McMurray, C. T.; Wilson, W. D. *FEBS Lett.* **1984**, *176*, 389.

65. den Hartog; J. H. J., Altona, C.; van Boom, J. H.; Reedijk, J. *FEBS Lett.* **1984**, *176*, 393.

66. Reily, M. D.; Marzilli, L. G. *J. Am. Chem. Soc.* **1985**, *107*, 4916.

67. Fouts, C. S.; Reily, M. D.; Marzilli, L. G.; Zon, G. *Inorg. Chim. Acta* **1987**, *137*, 1.

68. den Hartog, J. H. J.; Altona, C.; van der Marel, G. A.; Reedijk, J. *Eur. J. Biochem.* **1985**, *147*, 371.

69. den Hartog, J. H. J.; Altona, C.; Chottard, J.- C.; Girault, J.- P.; Lallemand, J.- Y.; de Leeuw, F. A. A. M.; Marcelis, A. T. A.; Reedijk, J. *Nucleic Acids Res.* **1982**, *10*, 4715.

70. van der Veer, J. L.; van der Marel, G. A.; van den Elst, H.; Reedijk, J. *Inorg. Chem.* **1987**, *26*, 2272.

71. Fouts, C. S.; Marzilli, L. G.; Byrd, R. A.; Summers, M. F.; Zon, G.; Shinozuka, K. *Inorg. Chem.* **1988**, *27*, 366.

72. Caradonna, J. P.; Lippard, S. J. *Inorg. Chem.* **1988**, *27*, 1454.

73. Sherman, S. E.; Gibson, D.; Wang, A. H. -J.; Lippard, S. J. *Science* **1985**, *230*, 412.

74. Admiraal, G. A.; van der Veer, J. L.; de Graaff, R. A. G.; den Hartog, J. H. J.; Reedijk, J. *J. Am. Chem. Soc.* **1987**, *109*, 592.

75. Fouts, C. S. Ph.D. Thesis, Emory University, Atlanta, GA, **1987**.

76. Marzilli, L. G.; Fouts, C. S.; Kline, T. P.; Zon, G. In *Platinum and Other Metal Coordination Compounds in Cancer Chemotherapy*; M. Nicolini, Ed., Martinus Nijhoff Publishing: Boston, **1987**; p. 67.

77. den Hartog, J. H. J.; Altona, C.; van Boom, J. H.; van der Marel, G.; Haasnoot, C. A. G.; Reedijk, J. *J. Biomol. Struct. Dyn.* **1985**, *2*, 1137.

78. Marcelis, A. T. M.; den Hartog, J. H. J.; van der Marel, G.; Wille, G.; Reedijk, J. *Eur. J. Biochem.* **1983**, *135*, 343.

79. Tran-Dinh, S.; Neumann, J.- M.; Huynh-Dinh, T.; Igolen, J.; Kan, S. K. *Org. Mag. Res.* **1982**, *18*, 148.

80. Neumann, J.- M.; Tran-Dinh, S.; Girault, J.- P.; Chottard, J.- C.; Huynh-Dinh, T. *Eur. J. Biochem.* **1984**, *141*, 465.

81. Sarma, R. H., In *Nucleic Acid Geometry and Dynamics*; R. H. Sarma, Ed., Pergamon Press, Inc.: New York, **1980**.

82. Eadie, J. S.; McBride, L. J.; Efcavitch, J. W.; Hoff, L. B.; Cathcart, R. *Anal. Chem.* **1987**, *165*, 442.

83. Basu, A. K.; Essigman, J. M. *Chem. Res. Toxicol.* **1988**, *1*, 1.

84. Lankhorst, P. P.; Haasnoot, C. A. G.; Erkelens, C.; Altona, C. *J. Biomol. Struct. Dyn.* **1984**, *1*, 1387.

85. Miller, S. K.; Marzilli, L. G. *Inorg. Chem.* **1985**, *24*, 2421.

86. Hollis, L. S.; Stern, E. W.; Amundsen, A. R.; Miller, A. V.; Doran, S. L. *J. Am. Chem. Soc.* **1987**, *109*, 3596.

87. Kozelka, J.; Petsko, G. A.; Quigley, G. J.; Lippard, S. J. *Inorg. Chem.* **1986**, *25*, 1075.

88. Rice, J. A.; Crothers, D. M.; Pinto, A. L.; Lippard, S. J. *Proc. Natl. Acad. Sci., USA* **1988**, *85*, 4158.

RECEIVED May 11, 1989

Chapter 10

Base-Selective DNA Cleavage with a Cyclometalated Palladium Complex

J. William Suggs, John D. Higgins III, Richard W. Wagner, and Julie T. Millard

Department of Chemistry, Brown University, Providence, RI 02912

The cyclometalated complex chloro(N,N,6C')-2(2-methoxy-phenyl)-1,10-phenanthroline palladium, 1, in the presence of H_2O_2, causes DNA strand scission at dG. The 5' end of the resulting nick is a phosphate, while the 3' end appears to be chemically heterogeneous, but is not a 3'-phosphate. Neither 1 alone, H_2O_2 alone, nor Li_2PdCl_4 plus H_2O_2 gives dG cleavage. Other cyclometalated palladium complexes, such as the one derived from azobenzene, showed no DNA nicking.

A NUMBER OF TRANSITION METAL-CONTAINING COMPOUNDS exhibit interesting DNA processing properties, including selective DNA nicking activities. One such reagent is bis(1,10-phenanthroline)Cu(I) [Cu(OP)$_2$] which, in the presence of hydrogen peroxide, exhibits powerful nucleolytic activity (*1*). Cu(OP)$_2$ was shown to cleave with high selectivity the unusual dinucleotide repeat conformation adopted by alternating d(AT)$_n$ sequences (*2*). The reasons for this selectivity are not firmly established, nor it is known with certainty the molecular species which is responsible for DNA cleavage in this system. The ability of Cu(OP)$_2$ to cleave tracts of d(AT)$_n$ sequences with little spillover onto adjacent sequences seems most consistent with the active species being some copper-oxo reagent rather than the diffusible hydroxyl radical.

We investigated other complexes of redox active metals to find out if they, like Cu(OP)$_2$, recognized d(AT)$_n$ regions. Among transition metal complexes, hundreds of structurally diverse cyclometallated palladium complexes have been synthesized. These are complexes in which a ligand atom (such as N, P, O, or S) directs a transition metal to metallate a nearby carbon, producing a five- (or occasionally four- or six-) membered chelate (*3*). Interestingly, there is a report of DNA nicking activity by one of these palladium-carbon bond containing species (*4*). Nevertheless, the selectivity, mechanism and generality of DNA nicking by these palladium

0097–6156/89/0402–00146$06.00/0

complexes is completely unexplored. Thus, we chose to compare the nuclease activities of a few of these cyclometallated compounds with the Cu(OP)$_2$ reagent. Unexpectedly, some of the palladium complexes gave specific cleavage at dG residues rather than showing selectivity for d(AT)$_n$ tracts. The complex with greatest activity, among the limited number we have surveyed, was chloro(N,N,6C´)-2-(2-methoxyphenyl)-1,10-phenanthroline palladium (1).

DNA Binding

The cyclometallated palladium complex **1** was prepared in our lab in the course of another project. It was synthesized via the addition of 2-lithio anisole to 1,10-phenanthroline followed by HgO oxidation to give the organic ligand. This was cyclometallated in MeOH with Li$_2$PdCl$_4$. Overall, the yield of **1** was over 30%, with the cyclometallation reaction essentially quantitative. A single crystal X-ray diffraction study of a derivative of **1**, in which the chloride had been replaced by a tert-butyl S-sulfinate group showed all four rings to be coplanar within 2° (J. W. Suggs

1

and J. D. Higgins, III, manuscript in preparation). Structurally, **1** is similar to the terpyridyl complexes investigated by Lippard's group (5), to the extent that a planar ligand occupies three of four coordination sites on a square planar metal. Since one of the sites in **1** is a carbon, the complex is neutral unless ionization of the Pd–Cl bond takes place. In fact, analogous complexes are known to ionize in aqueous solution (6). The solubility of **1** in 50 mM Tris•HCl, pH 8.0, at 25 °C was measured as approximately 0.075 mg/ml (180 μM). These solutions obeyed Beer's law and centrifugation or filtration through 0.2 micron filters did not change the absorbance.

The structure of **1** suggests that it should act as a metallointercalator. Thus, competition DNA binding experiments with the intercalator ethidium bromide were carried out. Successive titrations of ethidium bromide into a solution of **1** and calf thymus DNA were followed by fluorescence spectroscopy of the ethidium bromide. The values which were obtained from a fluorescent Scatchard plot are given in Table I, using 9.24 μM calf thymus DNA and a nucleotide:1 ratio of 5:1. The slope of the Scatchard plot was 5×10^5 and the intercept (the apparent number of binding sites per base pair) was 0.16. As the concentration of **1** was increased to nucleotide:1 ratios of 2.5:1 and 1:1 both the slope and the intercepts changed to 4×10^5 and 0.10 and 3.2×10^5 and 0.04, respectively. This behavior, in which the slopes and intercepts change in intercalator competition experiments, is thought to arise when the competing ligand binds both as an intercalator and covalently (7). The covalent interaction (probably between the metal and a donor site such as N-7 of purines) could block intercalation at nearby sites producing an apparent decrease in available binding sites.

Table I. Fluorescence Scatchard Plot Parameters for the Binding of Ethidium Bromide to Calf Thymus DNA in the Presence of 1 (Molecular Ratio 5:1 DNA:1)

[EtBr](μM)	I/I_{max}	C_B(μM)	C_F(μM)	r	r/C_F(μM^{-1})
0.37	0.45	0.15	0.22	0.016	73×10^{-3}
0.73	0.81	0.27	0.46	0.024	63×10^{-3}
1.09	1.14	0.38	0.71	0.041	58×10^{-3}
1.45	1.41	0.47	0.98	0.051	52×10^{-3}
1.80	1.66	0.55	1.25	0.060	48×10^{-3}
2.16	1.89	0.62	1.54	0.067	44×10^{-3}
2.51	2.08	0.69	1.82	0.075	41×10^{-3}
2.85	2.27	0.75	2.10	0.081	39×10^{-3}
3.20	2.45	0.81	2.39	0.088	37×10^{-3}
3.54	2.81	0.93	2.61	0.101	38×10^{-3}

The data in Table I could be used to obtain a binding constant for **1** to calf thymus DNA. If C_M (the concentration of free metal complex) is constant for a given binding isotherm, equation (1) holds:

$$r_{Etd}/C_F = (n - r_{Etd})[K_{Etd}/(1 + K_M C_M)] \qquad (1)$$

where n = 0.2 (the maximum value of $r_{Etd} = 4 \times 10^5$ M^{-1} and r_{Etd} is the previously determined r. Solving for r_M, the ratio of bound metal complex per nucleotide and using equation (2),

$$C_M = [M]_{total} - r_M C_N \qquad (2)$$

one can calculate C_M, the concentration of free metal complex, using equation (3):

$$(n - r_{Etd}) K_{Etd} (C_F/r_{Etd}) - 1 = r_M[(1 + K_{Etd}C_F)/(n - r_M)] \quad (3)$$

Each data point in the 5:1 DNA:1 isotherm was used to yield a value for the apparent binding constant of **1** to calf thymus DNA, K_M, for each C_F. These are summarized in Table II. The average value for K_M was 1.0×10^5 M^{-1}. The same procedure applied to the 2.5:1 DNA:1 run (which exhibited a lower available binding site per nucleotide) gave an average K_M of 4.2×10^5 M^{-1}. These binding constants are similar to those for [(terpy)PtCl]$^+$ and [(terpy)PdCl]$^+$ (1.3×10^5 M^{-1} and 1.9×10^5 M^{-1}, respectively) which were measured in Lippard's group by this competition method (7).

Table II. Calculation of the Binding Constant K_M of **1** to Calf Thymus DNA from Each Point in Table I

r	$r/C_F(\mu M^{-1})$	r_M	$C_F(\mu M)$	C_M	$K_M(M^{-1})$
0.016	73×10^{-3}	0.0014	0.22	--	--
0.029	63×10^{-3}	0.0134	0.46	1.72×10^{-6}	5.0×10^{-4}
0.041	58×10^{-3}	0.014	0.71	1.72×10^{-6}	5.6×10^{-4}
0.051	52×10^{-3}	0.014	0.98	1.67×10^{-6}	8.7×10^{-4}
0.060	48×10^{-3}	0.021	1.25	1.66×10^{-6}	10.1×10^{-4}
0.067	44×10^{-3}	0.023	1.54	1.04×10^{-6}	12.7×10^{-4}
0.075	41×10^{-3}	0.023	1.82	1.64×10^{-6}	13.4×10^{-4}
0.081	39×10^{-3}	0.021	2.10	1.65×10^{-6}	13.4×10^{-4}
0.088	37×10^{-3}	0.014	2.39	1.67×10^{-6}	12.6×10^{-4}
0.101	38×10^{-3}	0.004	2.61	--	--

Upon excitation of **1** at 320 nm in water, a weak fluorescence maximum at 410 nm was observed. Addition of nucleic acids enhanced the fluorescence. The ordering of this enhancement was poly(dG-dC)•poly(dG-dC) > calf thymus DNA > tRNA (*E. coli*) > rRNA (bovine liver). No enhancement was observed with poly(dA-dT)•poly(dA-dT), or poly A. Many intercalators exhibit fluorescence when layered into d(AT)-rich DNA, but have their fluorescence quenched by d(GC) base pairs (8). Other intercalators, such as ethidium bromide, fluoresce equally well in any DNA sequence. However, we know of no example of an intercalator whose fluorescence is enhanced by d(GC) relative to d(AT) base pairs. It is more likely that the observed fluorescence enhancement arises from the fraction of the complexes which bind covalently to dG residues in DNA. However, enhancement was not directly proportional to the d(GC) content of different DNAs.

Several cyclometallated complexes, in addition to **1**, were investigated for possible covalent binding to closed circular DNA, using an agarose gel assay. [(terpy)PtCl]+, which, like **1** in the ethidium bromide competition assay, bound to calf thymus DNA intercalatively and covalently has been reported to retard the migration of closed circular DNA in 1% agarose gels (7). The cyclometallated compounds **1–5** were dissolved in N,N-dimethyl-

2   **3**   **4**

5

formamide and added to supercoiled plasmid pUC8 at 37 °C for 15 minutes (2 µM complex concentration, reagent/nucleotide ratio of 0.06, 10 v/v DMF/50 mM Tris•Cl, pH 8.0). Only complex **1** produced any retardation in the gel mobility. The failure of the platinum complex **5** to affect the gel mobility of pUC8 under these conditions is due to the relative kinetic inertness of platinum vs. palladium complexes since overnight incubation of **5** and pUC8 resulted in gel retardation identical to that induced by **1**.

The covalent binding of **1** to DNA was reasonably rapid. Incubation of pUC8 at 37 °C in a 10 µM solution of **1** for 3 minutes significantly changed the agarose gel mobility. As the concentrations of **1** increased to 100 µM, less and less of the nucleic acid was visible on the gel. This could result from **1** interfering with the staining of the DNA by ethidium bromide or from **1** acting to precipitate the DNA. The latter effect seems to dominate since addition of higher concentrations of **1** to ^{32}P-end labeled DNA restriction fragments, followed by centrifugation in a desk top microcentrifuge, pellets a portion of the DNA. Sodium cyanide reversed the binding of **1** to DNA. The rapidity of this reversal (less than 10 minutes) is another consequence of the kinetic lability of palladium–chloride bonds. Palladium cyanide bonds are not expected to ionize in aqueous solution, and therefore no vacant coordination site is available for DNA binding.

DNA Nicking with H_2O_2 and 1

Since the preceding experiments, particularly the fluorescence results, indicated that 1 binds to DNA with some selectivity, we next examined its ability to nick DNA in the presence of hydrogen peroxide, as the bis(1,10-phenanthroline)Cu/H_2O_2 system does. Both copper and palladium are redox-active metals. Palladium-based systems are often used as oxidants in organic chemistry. The Wacker process is one well-studied example. A filter binding assay (9) was used to follow the DNA nicking activity of 1.
[32]P-labeled alternating poly[d(A-T)] was prepared from [α–[32]P]dATP and dTTP. Reactions were run at 37 °C in TE buffer with 0.37% H_2O_2 and were 200 µM in 1. The molar ratios of 1 to nucleotide that we used are summarized in Table III. The times in Table III are the times after which no further DNA cleavage (as measured by filter binding) took place. Aliquots were removed and added to 5 ml of ice cold 5% trichloroacetic acid (TCA) with 1 mg/ml carrier calf thymus DNA. After 5 min, precipitated DNA was collected on a Whatman GF/C filter and washed twice with cold 10% TCA. The disks were dried and Cerenkov-counted. As a control reaction, Mung Bean Nuclease (10) gave complete solubilization with this assay after a four hour digestion. DNA fragments shorter than 10–15 base pairs are expected to be acid soluble.

Table III. Degree of Acid Solubilization of Various DNAs Upon Treatment With 1-H_2O_2 (0.3%)

DNA	1/base pair	time (hr)	% acid solubilization
poly[d(A-T)][a]	1:1	72	23
poly[d(A-T)]	1:1 (3% H_2O_2)	72	40
poly[d(A-T)]	2:1	50	49
poly[d(A-T)]	10:1	72	68
calf thymus[b]	0	72	1
calf thymus	1:1	72	1
calf thymus	10:1	72	16
calf thymus	10:1 (3% H_2O_2)	72	68
[32]P(dC) poly[d(G-C)[	10:1 (3% H_2O_2)	24	12
[32]P(dC) poly[d(G-C)]	20:1	24	18
[3]H(dG) poly[d(G-C)]	20:1	24	37
[32]P(dG) poly[d(G-C)]	20:1	24	48

[a]labeled with [^{32}P]dATP
[b]labeled with [^{32}P]dGTP

Sonicated calf thymus DNA was labeled under nick translation conditions with $[\alpha-^{32}P]dATP$ and the same filter binding assays were performed. After a 72 hour incubation with 1:10 DNA:1 molar ratio with 3.0% H_2O_2, 68% of the radioactivity was solubilized. Digestion of an aliquot of the starting DNA sample with an excess of DNase I for three hours gave a limit TCA solubilization of 65% of the total counts. These results show that the $1-H_2O_2$ combination, unlike bis(1,10-phenanthroline)Cu(I)/H_2O_2, is not an active nonspecific nuclease. For all the different DNAs used, hydrogen peroxide in the absence of 1 produced essentially no strand breaks, as measured by the filter binding assay.

Finally, alternating poly[d(G-C)] was nick translated with either $[\alpha-^{32}P]dCTP$, $[\alpha-^{32}P]dGTP$, or $[^3H-8]dGTP$. The results of filter binding assays are summarized in Table III. More solubilization was evident when the radioactive label was associated with a dG nucleotide than with a dC nucleotide. This is consistent with purine residues being much more reactive toward the $1-H_2O_2$ reagent than pyrimidine residues.

No basic pH treatment was needed in order to give DNA strand cleavage with $1-H_2O_2$. In order to determine the effect of pH on the degree of cleavage, the $[\alpha-^{32}P]dGTP$-labeled alternating poly(dG-dC) polymer was incubated with a 20:1 1:DNA (0.3% H_2O_2) reagent for 24 hr. Essentially identical amounts of acid soluble DNA were produced at pH 4.4 in a sodium phosphate buffer and at pH 7.5 in Tris buffer. However, the extent of cleavage was greatly enhanced (to 90%) in a pH 9.2 phosphate buffer. Likewise, treatment with piperidine following $1-H_2O_2$ at pH 7.5 produced 85% solubilization.

Several experiments were performed to characterize the functional groups produced during the DNA cleavage reaction. Since inorganic phosphate, unlike nucleotides, is not adsorbed by activated charcoal (11), the ^{32}P-labelled alternating d(GC) polymer reaction described above was stirred for 30 min with 1 g of activated charcoal. Upon filtration, all of the radioactivity was retained upon the filter. Thus, no inorganic phosphate is released during DNA strand cleavage by $1-H_2O_2$.

The presence of 3'- and 5'-terminal phosphate groups were assayed by comparing the release of ^{32}P following treatment with alkaline phosphatase vs. exonuclease III (12). Alkaline phosphatase removes both 3'- and 5'-phosphate groups; exonuclease III removes only 3'-phosphate groups. Typical $1-H_2O_2$ reaction conditions were carried out with ^{32}P-labelled poly(dG-dC) and incubated overnight. Reactions were set up in duplicate using an aliquot from the incubation and either water (control), exonuclease III (200 units, 37 °C, 30 min), or bacterial alkaline phosphatase (15 min, 37 °C, 12 units, 15 min at 56 °C, repeat with 12 more units of enzyme). To each reaction was added activated charcoal, followed by filtration. The control showed no release of inorganic phosphate, nor did the exonuclease III reaction. However, alkaline phosphatase did produce 16% solubilization of the total ^{32}P. As a further control, ^{32}P labelled DNA was added to the exonuclease III reaction, and radioactivity was released.

This indicates that the enzyme was not being inhibited by some species in the cleavage reaction mixture. These results (which are consistent with the sequencing gel results to be described below) indicate that 1-H_2O_2 produces no 3'-phosphoryl termini, while some 5'-phosphoryl termini form. In this respect, the 1-H_2O_2 cleavage is similar to that produced by bleomycin (13). However, unlike the bleomycin reaction, no base propenals or malondialdehyde-like compounds were released (14), as indicated by a negative color test with thiobarbituric acid (15).

In order to determine what kinds of small molecules were released, paper chromatography (16) was run on the 1-H_2O_2 reaction with poly(dG-dC) labelled with either [α-^{32}P]dGTP or [^{3}H-8]dGTP. In each case the radioactivity migrated with dGMP. Essentially no radioactivity was found migrating with guanine or deoxyguanosine. While guanine could be released and further degraded, no major breakdown products were observed by chromatography. More definitive HPLC analysis is underway to characterize the molecules released from DNA in this reaction.

In order to more fully establish the selectivity of the 1-H_2O_2 DNA nicking reaction, products from reaction with a 192 base pair HindIII/HaeII restriction fragment of the plasmid pRWAT11.1 (2) were examined on a sequencing gel. Strands were ^{32}P-labeled at either one of the 5´ ends or 3´ ends and incubated with 100 μM 1 and 0.03% H_2O_2 for 15 min at 37 °C. Initially a 2.5 mM NaCN quench was used to ensure that any DNA strands were demetallated. However, we subsequently found that this was unnecessary (probably due to the formamide loading buffer). An autoradiogram of the subsequent gel is shown in Figure 1. Unlike the Cu(I) reagent, which cleaved preferentially at the d(AT)$_{11}$ run of the restriction fragment, the cyclometallated palladium nuclease 1 cleaved selectively at dG residues. The 3´-end labeled strand migrated with a Maxam-Gilbert A+G lane, indicating that the 5´ end of the nicked strand is a phosphate. The 5´-end labeled strand also gave to a first approximation a dG cleavage pattern, although there was reproducibly considerable background in the lane and the shorter fragments had reduced electrophoretic mobility by 2–3 bases, compared to the Maxam-Gilbert lane. The probable reason for the muddy 5´-end labeled lane is chemical heterogenity of the 3´ ends of the fragments.

Control experiments with 1 alone, H_2O_2 alone, or Li$_2$PdCl$_4$ with or without H_2O_2, gave no sequence-selective nicking at dG residues. At pH 2, Li$_2$PdCl$_4$ specifically cleaves DNA at dA sites (17). However, this chemistry is absent at pH 8, where our experiments were done. DNA cleavage required an ordered addition of reagents. If H_2O_2 was added, followed by 1 and a rapid workup (< 1 minute) no dG lane resulted. Furthermore, when 1 and DNA were incubated, followed by trapping of 1 with 2.5 mM NaCN and addition of H_2O_2, only weak, nonspecific levels of DNA cleavage took place. Other cyclometallated complexes, in particular 2, 3, and 4, showed no general or dG-specific cleavage in the presence of H_2O_2 under the same reaction conditions as 1 (autoradiogram not shown).

Figure 1. Autoradiogram of an 8% polyacrylamide gel of the HindIII-HaeII fragment of the plasmid pRWAT 11.1. The ends were labeled 5′ or 3′ with ^{32}P (2). Lanes 1, 4: Maxam-Gilbert (A+G) lanes for 3′ and 5′ labeled DNA. Lanes 2 and 5: 100 μM 1 and 0.03% H_2O_2 (15 min, 100 μl total volume, 37 °C, 1 μg sonicated calf thymus DNA), for 3′ and 5′ labeled strands. Lane 3: Cu(OP)$_2$ for the 5′ strand. Lane 6: 0.03% H_2O_2 alone. Lane 7: 50 μM Li$_2$PdCl$_4$ and 0.03% H_2O_2. Lane 8: 100 μM 1, no H_2O_2.

The conditions which produced the dG-specific results were repeated on the 163 base pair 3′-end labeled HindIII/HaeII fragment of pUC8. Densitometry (*18*) of the resulting sequencing gel autoradiogram allowed the calculation of the normalized probability for cutting at each site in the central region of the restriction fragment, which is plotted on a logarithmic scale in Figure 2. For comparison, a densitometer trace of a Maxam-Gilbert dG-specific reaction (dimethyl sulfate-piperidine run slightly longer than normal) was prepared.

It is interesting that both traces show dG specificity. Nevertheless, some inappropriate cleavages are seen as well, and these mistakes are the same for both reagents. There are weak dA hits at positions 112, 127, 137, and 145, a strong dA cleavage at 112, and a strong dT hit at 123. Also, the first dG at 117 in a daGGG sequence is resistant to cleavage by 1-H_2O_2.

Binding of **1** to DNA does not cause strand separation, since treatment of the 3′-^{32}P-labeled 163 base pair fragment incubated with **1** followed by S1 nuclease gave no S1 hotspots. Further, unlike bis(1,10-phenanthroline)Cu(I), **1** does not bind in the DNA minor groove. Anthramycin, which binds covalently to the dG-NH_2 group in the minor groove (*19*) was preincubated with the 3′-^{32}P-labeled 163 base pair fragment. Subsequent nicking with 1-H_2O_2 caused no significant change in the dG ladder.

Discussion

A number of small molecule agents have been shown to produce single-strand breaks under oxidative conditions in DNA. Some, such as singlet oxygen, modify one or more DNA bases, leading to strand scission (*20*). More often, metal-containing species, including Fe(II)EDTA, Cu(OP)$_2$, and bleomycin, act to abstract hydrogen atoms from the deoxyribose ring, which initiates strand breakage. The cyclometallated palladium nuclease **1** shows both similarities and differences to other metal-based DNA nicking agents.

Most interestingly, 1-H_2O_2 shows excellent base selectivity, producing strand breaks at dG residues. The ethidium bromide competition binding data indicate that, like the structurally related [(terpy)PtCl]$^+$, [(terpy)PdCl]$^+$, and (en)PdCl$_2$ complexes, **1** binds both intercalatively and covalently to DNA. Intercalative binding is not, by itself, sufficient to account for the dG-selective nicking of **1**, since intercalators have only a limited preference for 5′-pyrimidine-purine-3′ sequences (*21*). It is probable that the covalent binding of **1** to DNA is responsible for the dG selectivity. Terpyridyl and related ligands are known to labilize chloride ions, and the ability of cyanide to block nicking indicates that ionization of the Pd–Cl bond in **1**, which generates a vacant covalent binding site on the metal, plays an important role in the DNA binding properties of **1**. The N7 position of guanine is the favored platinum group metal binding site in

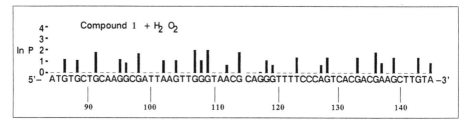

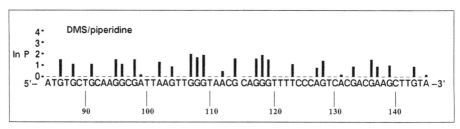

Figure 2. Densitometer trace within the 163 base pair HindIII/HaeII fragment of pUC8 showing the cleavage specificity of 1 in the presence of H_2O_2 and the Maxam-Gilbert dG reaction. The numbering scheme is base pairs from the 5-end of the fragment.

DNA (22) and should be the the covalent binding site for 1 upon ionization.

Like bleomycin, 1-H_2O_2 produces both nicks in DNA and alkali-labile lesions (23). However, the cyclometallated nuclease does not produce malondialdehyde or related thiobarbituric acid-reactive functional groups and free DNA bases are not detected as reaction products. In order to understand the mechanism by which 1 nicks DNA, it will be necessary to examine the cleavage products of short oligonucleotides, a strategy which has proven useful in understanding the mode of action of bleomycin.

Given the hundreds of cyclometallated palladium complexes which have been described in the literature, and the predictable manner in which they can be formed, it is likely that more efficient and selective cyclometallated nucleases can be designed. Since essentially all the transition elements can form air- and water-stable cyclometallated complexes, both the metal and ligand can be varied to produce new DNA nicking agents related to 1.

Experimental

The procedures involved in DNA isolation, purification, labeling, and autoradiography were as previously reported (2). The synthesis and characterization of 1 will be fully described elsewhere (J. W. Suggs and J. D. Higgins, III, manuscript in preparation). Fluorescence measurements were made on a Gilford Fluoro IV spectrofluorometer at 25 °C. The excitation wavelength for the ethidium bromide competition experiments was 540 nm; emission, 650 nm. Competition studies were carried out at successive titration with ethidium bromide into a solution of 1 and calf thymus DNA in 50 mM Tris•HCl, 0.2 M NaCl, pH 7.5. The concentration of bound ethidium bromide, C_B, was calculated from the fluorescence intensity I by the equation $C_B=(I/I_{max})[EtdBr]_{sat}$, where $[EtdBr]_{sat}$ was a small known concentration in the presence of saturating amounts of DNA, and I_{max} was the intensity in that case. C_F (free ethidium bromide) was obtained by subtracting C_B from the total ethidium bromide concentration, C_T. The ratio (r) of bound ethidium to total nucleotide concentration was calculated from $r=C_B/C_{No}$ where C_{No} is the concentration of base pairs.

Filter binding assays were performed on 100 μl samples in Tris buffer, pH 7.5, which were 2 mM in DNA, 0.3% by volume hydrogen peroxide, and contained approximately 10^5 cpm (Cerenkov) of ^{32}P-labeled DNA. The label was introduced by nick translation. Aliquots were precipitated with 5% trichloroacetic acid onto Whatman GF-C glass filters. The filters were washed with 5% TCA and ethanol, air dried and Cerenkov-counted.

In order to check for the release of inorganic phosphate, ^{32}P labeled alternating poly[d(G-C)] was incubated as above with 1 (DNA:1 1:20) for 24 hours with 0.3% hydrogen peroxide. Then 3 ml of 5 mM Tris (pH 7.8) and 1 g of activated charcoal were added and the mixture vortexed, followed by chilling at 0 °C for 30 min. Filtration through a glass fiber filter led to only 1% of the radioactivity in the filtrate.

Acknowledgments

This work was supported by the Public Health Service. We thank Dr. A. Batcho for a gift of anthramycin. J.W.S. is a Research Career Development Awardee of the NCI, CA00947.

Literature Cited

1. Hecht, S. M. *Acc. Chem. Res.* **1986**, *19*, 383–391.
2. Suggs, J. W.; Wagner, R. W. *Nucleic Acids Res.* **1986**, *14*, 3703–3716.
3. Newkome, G. R.; Puckett, W. E.; Gupta, V. K.; Kiefer, G. E. *Chem. Rev.* **1986**, *86*, 451–489.
4. Newkome, G. R.; Onishi, M.; Puckett, W. E.; Deutsch, W. A. *J. Am. Chem. Soc.* **1980**, *102*, 4551–4552.
5. Lippard, S. J. *Acc. Chem. Res.* **1978**, *11*, 211–217.
6. Basolo, F.; Gray, H. B.; Pearson, R. G. *J. Am. Chem. Soc.* **1960**, *82*, 4200–4203.
7. Howe-Grant, M.; Wu, K. C.; Bauer, W. R.; Lippard, S. J. *Biochemistry* **1976**, *15*, 4339–4346.
8. Waring, M. J. *J. Mol. Biol.* **1965**, *13*, 269–282.
9. Moses, R.E. In *Methods in Enzymology*; Colowick, S. P.; Kaplan, N. O. Ed.; Academic : New York, **1972**, Vol. 29, p. 13.
10. Johnson, P. H.; Laskowski, M., Sr. *J. Biol. Chem.* **1970**, *245*, 891–898.
11. Smith, M.; Khorana, H. G. In *Methods in Enzymology*; Colowick, S. P.; Kaplan, N. O. Ed.; Academic: New York, **1963**, Vol. 6, p. 659.
12. Kuo, M. T.; Haidle, C. W. *Biochem. Biophys. Acta* **1973**, *335*, 109–114.
13. Giloni, L.; Takeshita, M.; Johnson, F.; Iden, C.; Grollman, A. P. *J. Biol. Chem.* **1981**, *256*, 8606–8615.
14. Burger, R. M.; Berkowitz, A. R.; Peisach, J.; Horwitz, S. B. *J. Biol. Chem.* **1980**, *255*, 11832–11838.
15. Waravdekar, V. S.; Saslaw, L. D. *J. Biol. Chem.* **1959**, *234*, 1945–1950.
16. Rhaese, J.-J.; Freese, E. *Biochem. Biophys. Acta* **1968**, *155*, 476–490.
17. Iverson, B. L.; Dervan, P. B. *Nucleic Acids Res.* **1987**, *15*, 7823–7830.
18. Lutter, L. C. *J. Mol. Biol.* **1978**, *124*, 391–420.
19. Hurley, L. H.; Thurston, D. E. *Pharm. Res.* **1984**, 52–67.
20. Friedmann, T.; Brown, D. M. *Nucleic Acids Res.* **1978**, *5*, 615–622.
21. Krugh, T. R.; Reinhardt, C. G. *J. Mol. Biol.* **1975**, *97*, 133–162.
22. Sherman, S. E.; Gibson, D.; Wang, A. H.-J.; Lippard, S. J. *Science* **1985**, *230*, 412–417.
23. Sugiyama, H.; Xu, C.; Murugesan, N.; Hecht, S. M.; van der Marel, G. A.; van Boom, J. H. *Biochemistry* **1988**, *27*, 58–67.

RECEIVED January 26, 1989

Chapter 11

Metal–Nucleotide Interactions

Helmut Sigel

University of Basel, Institute of Inorganic Chemistry, Spitalstrasse 51,
CH–4056 Basel, Switzerland

Nucleotides and their metal ion complexes undergo self-association via base-stacking in aqueous solution. The extent of this association is quantified for several systems, and conditions are given which should be used in studies directed toward determination of the properties of the monomeric species. Stability constants determined by potentiometric pH titrations for the complexes between Mg^{2+}, Ca^{2+}, Mn^{2+}, Co^{2+}, Ni^{2+}, Cu^{2+}, Zn^{2+} or Cd^{2+} and adenosine 5'-triphosphate (ATP^{4-}) and the three pyrimidine nucleoside 5'-triphosphates are listed; evaluation of stability differences allows calculation of the degree of formation of the macrochelated N-7 back-bound species in $M(ATP)^{2-}$ complexes. There is evidence (including NMR and ultraviolet perturbation studies) that coordination to N-7 may occur either directly or outer sphere, i.e. via a coordinated water molecule. For monophosphate esters (R-MP^{2-}) the relationship between complex stability and phosphate group basicity is established. The extent of the nucleic base-metal ion interaction is quantified for the M(NMP) species of several nucleoside monophosphates (NMP^{2-}) including pyrimidine and purine derivatives (e.g., 2'-AMP^{2-}, 3'-AMP^{2-}, 5'-AMP^{2-}, and 5'-GMP^{2-}). The properties of several mixed ligand complexes, as well as some conclusions relating to interactions between metal ions and nucleic acids, are indicated.

Nucleotides and metal ions are involved in the basic metabolic processes of life (1–3). Since Szent-Györgyi (4) proposed in 1956 a macrochelate for the complex formed between adenosine 5'-triphosphate (ATP) and Mg^{2+}, this idea has fascinated many chemists and molecular biologists. Today we know that the binding sites of ATP are only in part those originally suggested (4). Many of the early advances were made by

0097–6156/89/0402–00159$12.25/0

© 1989 American Chemical Society

the groups of Cohn (5, 6), Brintzinger (7–10), and Shulman (11, 12): the stabilities of the complexes formed with the alkaline earth ions and the second half of the divalent 3d transition metal ions are largely determined by the coordination tendency of the phosphate groups; interaction of a phosphate-coordinated metal ion with an adenine residue occurs (if at all) via N-7. It was also soon discovered that this latter interaction, which is responsible for the formation of $M(ATP)^{2-}$ macrochelates, can be inhibited by the formation of mixed-ligand complexes, e.g. with 2,2'-bipyridyl (13–15).

Further progress was hampered by the fact that self-association of nucleotides and their complexes was not yet recognized. Studies of ATP systems by UV and NMR measurements led to different conclusions (16). The latter measurements were carried out with concentrations typically on the order of 0.05 to 0.3 M, while in the former about 10^{-4} to 10^{-3} M solutions were employed. Today we know that by UV methods the properties of monomeric species are observed, while by NMR associated species are studied under the concentrations cited above. Therefore, in the next section some of the results regarding self-association of nucleotides shall be summarized briefly.

Self-Association of Nucleotide Systems

In the early sixties Ts'o and his coworkers established that in aqueous solution nucleic bases and nucleosides undergo self-association via base stacking (17,18). However, in the following years self-stacking of nucleotides was often ignored with the argument that repulsion by the negatively-charged phosphate groups would render self-association insignificant. This assumption is incorrect: certainly, repulsion reduces the self-association tendency of nucleotides if compared with the corresponding nucleosides, but in the presence of metal ions at least partial charge neutralization occurs via metal ion-phosphate coordination.

By ^{1}H NMR shift experiments self-association of nucleotides (N) can be demonstrated, and it is now generally accepted that self-association occurs via stacking and that oligomers are formed (19, 20). The experimental data of ^{1}H NMR measurements can be explained and quantified with the isodesmic model for an indefinite non-cooperative association (19, 21–24). In this model it is assumed that the equilibrium constants (Equation 2) for the equilibria in Equation 1 are all equal:

$$N_n + N \rightleftharpoons N_{n+1} \tag{1}$$

$$K = \frac{[N_{n+1}]}{([N_n][N])} \tag{2}$$

The self-stacking tendency decreases within the nucleoside series (24) adenosine ($K = 15 \pm 3$ M^{-1}) > guanosine (8 ± 3 M^{-1}) > inosine (3.3 ± 0.3

M^{-1}) > cytidine (1.4 ± 0.5 M^{-1}) ~ uridine (1.2 ± 0.5 M^{-1}); the same order applies for the corresponding 5'-nucleotides (24, 25).

Association constants (Equation 2) for several derivatives of adenine and uracil are compiled as examples in Table I (24–28). Many comparisons may be made; a few follow: (i) As one might expect, the tendency for self-stacking decreases in the series adenosine > AMP^{2-} > ADP^{3-} > ATP^{4-} (24, 25). (ii) The tendency for self-stacking of the adenosine monophosphates (AMP^{2-}) is not significantly dependent on the position of the phosphate group at the ribose moiety (27). (iii) The degree of protonation is important; e.g., with 5'-AMP self-association is at maximum, if the phosphate groups are monoprotonated and if in addition 50% of the adenine residues carry a proton at N-1 (26). (iv) Complete protonation at N-1 of adenosine and of 5'-AMP, giving $H(Ado)^+$ and $H_2(AMP)^{\pm}$, respectively, inhibits self-stacking strongly (26), whereas $H_2(5'\text{-ATP})^{2-}$ with a proton each at the terminal phosphate group and at N-1 of the adenine moiety has the largest self-association tendency indicating that here also ionic interactions are important (28). (v) Comparison of the self-association tendency of the isocharged species $H(5'\text{-AMP})^-$ and $Mg(5'\text{-ADP})^-$ or 5'-AMP^{2-} and $Mg(ATP)^{2-}$ indicates that the Mg^{2+} complexes self-associate more strongly; this observation has led to the suggestion (29) that bridging by Mg^{2+} *via* the phosphate residues in the oligomers may occur. However, it should be emphasized that the ^1H NMR data (24, 25) give no indication of a Mg^{2+}/base interaction in the stacks. (vi) This is different in $M(ADP)^-$ and $M(ATP)^{2-}$ systems (24, 25) with M = Zn^{2+} or Cd^{2+}; in these cases the large tendency for self-association is certainly linked to a M^{2+}/N-7 interaction favoring dimeric stacks (for details see 24, 25, 29).

As a guide for developing a feeling of how the proportions of the various oligomers (Equation 1) vary with changing concentration the results of Figure 1 are provided, which were calculated with K = 15 M^{-1}. It is evident that at concentrations greater than 0.01 M the degree of formation of oligomers is remarkable; more examples of such distribution curves are given in (24–29). This type of self-association is most probably of biological significance: for example, the concentration of ATP in the chromaffin granules of the adrenal medulla is about 0.15 M (30–32), and there are also substantial amounts of metal ions present. Hence, (self)-stacking interactions should be considerable, especially as the concentration of catecholamines (30–32) is also very large, allowing the formation of mixed stacks (see, e.g., 29, 33, 34).

Finally, it is important to obtain an impression of the concentration range in which monomeric species strongly dominate. Adenosine may be used as an example (K = 15 M^{-1}) (26, 27). In a 1 mM solution about 97% of the total adenosine exist in the monomeric form. Hence, no experiments aiming at the properties of monomeric nucleotide species should be carried out in concentrations higher than 10^{-3} M (35, 36). In fact, often it may be necessary to work even at lower concentrations; e.g., with

Table I. Association Constants for Self-Stacking (Equations 1,2) of
Several Adenine and Uracil Derivatives, Together with the Influence of
Protons (D[+]) or Metal Ions (M[2+]) as Determined by ^{1}H NMR Shift
Measurements in D_2O at 27 °C

System		I (M)[a]	K (M[-1])[b]	Ref
adenine derivatives				
adenosine (Ado)		0.1	15 ± 3	26
	Ado/D(Ado)[+] ≈ 1:1	0.1	6.0 ± 1.3	26
	D(Ado)[+]	0.1– ~0.4	0.9 ± 0.2	26
AMP:	2'-AMP[2-]	0.1– ~1.2	2.0 ± 0.2	26,27
	3'-AMP[2-]	0.1– ~1.2	1.6 ± 0.4	26,27
	5'-AMP[2-]	0.1– ~1.2	2.1 ± 0.3	26,27
	D(5'-AMP)[-]	0.1– ~0.6	3.4 ± 0.3	26
	D(5'-AMP)[-]/D_2(5'-AMP)[±] ≈ 1:1	0.1– ~0.6	5.6 ± 0.5	26
	D_2(5'-AMP)[±]/D_3(5'-AMP)[+] ≈ 3:1	0.1– ~0.2	0.9 ± 0.2	26
5'-ADP:	ADP[3-]	0.1– ~1.7	1.8 ± 0.5	25
	Mg(ADP)[-]	0.1– ~1.2	6.4 ± 0.9	25
	Zn(ADP)[-]	0.1– ~0.15	~100[c]	25
	Cd(ADP)[-]	0.1– ~0.15	~100[c]	25
5'-ATP:	ATP[4-]	0.1– ~2	1.3 ± 0.2	24,28
	D(ATP)[3-]	0.1– ~2	2.1 ± 0.3	28
	D(ATP)[3-]/D_2(ATP)[2-] ≈ 1:1	0.1– ~1.7	6.0 ± 2.0	28
	D_2(ATP)[2-]	0.1– ~0.7	~200[c]	28
	D_3(ATP)[-]/D_2(ATP)[2-] ≈ 4:1	0.1– ~0.35	17 ± 10[c]	28
	Mg(ATP)[2-]	0.1– ~2	4.0 ± 0.5	24
	Zn(ATP)[2-]	0.1– ~2	~11.1 ± 4.5[c]	24
	Cd(ATP)[2-]	0.1– ~2	~17[c]	24
uracil derivatives				
uridine		0.1	1.2 ± 0.5	24
5'-UDP:	UDP[3-]	0.1– ~1.7	0.6	25
	Mg(UDP)[-]	0.1– ~1.2	1.4	25
5'-UTP:	UTP[4-]	0.1– ~2	~0.4	24
	Mg(UTP)[2-]	0.1– ~0.5	d	24

[a]The ionic strength (I) was adjusted to 0.1 M by adding $NaNO_3$ where necessary.
[b]The error range is twice the standard error (2σ).
[c]These values are only estimates as the experimental data cannot be completely explained
   by the isodesmic model (Equations 1, 2); see details in (24, 25, 28).
[d]Only solutions up to [Mg[2+]] = [UTP] = 0.1 M could be studied due to precipitation, and the
   data for this limited concentration range could be fitted within experimental error with
   the value of K obtained for UTP[4-]. This indicates that the self-association tendency for
   Mg(UTP)[2-] is only slightly larger than for UTP[4-].

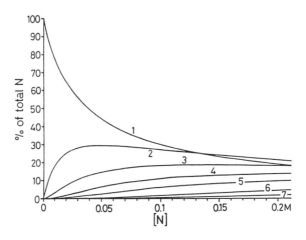

Figure 1. Variation of the proportions of a nucleoside derivative N present in the monomer (1), dimer (2), trimer (3), ... and heptamer (7) in D$_2$O solution as a function of the total concentration of N; calculated for a self-association constant K = 15 M^{-1} (see Table I).

Zn(ADP)⁻ concentrations below 0.3 mM are required (25). These pitfalls have been considered in the experiments which are summarized in the following sections.

Evidence for Purine N-7/Backbinding and Stability of Nucleoside 5'-Triphosphate Complexes

The ^{1}H NMR shift data mentioned above may easily be extrapolated to zero concentration of the nucleotide systems. In this way the chemical shifts of H-8 (Figure 2; *vide infra*) for ATP^{4-}, $Mg(ATP)^{2-}$, $Zn(ATP)^{2-}$, and $Cd(ATP)^{2-}$ are obtained for the monomeric species. The same is possible for the corresponding complexes with ITP^{4-} and GTP^{4-} (24). The result that is found is that the chemical shifts for each pair of a nucleoside 5'-triphosphate (NTP^{4-}) and its $Mg(NTP)^{2-}$ complex are very similar, with no indication for a downfield shift of H-8 in the Mg^{2+} complex. The result is different if the chemical shifts of $Zn(NTP)^{2-}$ and $Cd(NTP)^{2-}$ are compared with those of the corresponding $Mg(NTP)^{2-}$ (or NTP^{4-}): in all cases significant downfield shifts are observed, providing direct evidence that a M^{2+}/N-7 interaction occurs (24, 25). Hence, for any nucleoside phosphate (NP) complex of a purine derivative the intramolecular equilibrium (Equation 3) between an "open" species, $M(NP)_{op}$, and a "closed" species, $M(NP)_{cl}$, should be considered:

$$
\begin{array}{c}
\text{phosphate}\underset{\underset{\displaystyle M^{2+}}{=}}{=}\text{ribose}\underset{=}{=}\text{base}
\end{array}
\quad \underset{\rightleftharpoons}{K_I} \quad
\begin{array}{c}
\text{phosphate} - \text{r} \\
\underset{\underset{\underset{\displaystyle \text{base}\ -\ \text{e}}{\equiv}}{\equiv}}{M^{2+}}\ \ \begin{array}{l} \text{i} \\ \text{b} \\ \text{o} \\ \text{s} \end{array}
\end{array}
\qquad (3)
$$

In those cases where the equilibrium in Equation 3 exists and the closed species is formed to some extent, this will be reflected in an increased stability of the M(NP) complex (37).

The dimensionless intramolecular equilibrium constant K_I may be calculated (24, 37) via Equation 4,

$$
K_I = \frac{[M(NP)_{cl}]}{[M(NP)_{op}]} = \frac{K^M_{M(NP)}}{K^M_{M(NP)_{op}}} - 1 = 10^{\log \Delta} - 1 \qquad (4)
$$

where $\log \Delta$ is defined by Equation 5:

$$
\log \Delta = \log K^M_{M(NP)} - \log K^M_{M(NP)_{op}} \qquad (5)
$$

The remaining definitions are given in Equations 6 through 8:

$$M + NP \rightleftarrows M(NP)_{op} \rightleftarrows M(NP)_{cl} \qquad (6)$$

$$K^{M}_{M(NP)_{op}} = \frac{[M(NP)_{op}]}{[M][NP]} \qquad (7)$$

$$K^{M}_{M(NP)} = \frac{[M(NP)]}{[M][NP]} = \frac{([M(NP)_{op}] + [M(NP)_{cl}])}{[M][NP]} \qquad (8a)$$

$$K^{M}_{M(NP)} = K^{M}_{M(NP)_{op}} + K_I \cdot K^{M}_{M(NP)_{op}} = K^{M}_{M(NP)_{op}} (1 + K_I) \qquad (8b)$$

Equation 4 follows from Equation 8b. The overall stability constant $K^{M}_{M(NP)}$ (Equation 8a) is experimentally accessible; the difficulty is to obtain values for $K^{M}_{M(NP)_{op}}$ (Equation 7), as the stability of the open species cannot directly be measured. Furthermore, it is evident that the reliability of any calculation for K_I (Equation 4) depends on the accuracy of the difference $\log \Delta$ (Equation 5), and this accuracy depends on the error limits of the constants $K^{M}_{M(NP)}$ and $K^{M}_{M(NP)_{op}}$ which become more important the more similar the two constants are.

As discussed above, the ^1H NMR shift results show that the equilibrium in Equation 3 exists at least for some M^{2+} complexes of purine-nucleoside 5'-triphosphates (*24, 35, 36*). As ATP^{4-} occurs in aqueous solution mainly in the *anti* conformation (*27, 38, 39*), which is depicted in Figure 2A, a metal ion coordinated to the phosphate chain may also reach N-7. In the *anti* conformation the C-8/N-9 bond of purines (Figure 2) and the C-6/N-1 bond of pyrimidines projects onto or near the ribose ring (*38, 39*).

Regarding evaluation of the position of the equilibrium in Equation 3 for $M(ATP)^{2-}$ complexes, two favorable circumstances exist: (i) ^1H NMR shift studies (*24*) for pyrimidine-nucleoside 5'-triphosphates ($PNTP^{4-}$) with Mg^{2+}, Zn^{2+}, or Cd^{2+} show that the pyrimidine moiety of UTP^{4-}, TTP^{4-} and CTP^{4-} (Figure 2B) is not involved in complex formation. (ii) The proton basicity of the terminal γ-phosphate group for all the NTPs shown in Figure 2 is very similar; i.e., for the negative logarithm of the acidity constants it holds: $pK^{H}_{H(NTP)} = 6.50 \pm 0.05$ (I = 0.1 M, $NaNO_3$ or $NaClO_4$; 25 °C) (*36*). Hence, the stability constants $K^{M}_{M(NTP)}$ measured for the corresponding $M(NTP)^{2-}$ complexes may directly be compared, i.e. without an adjustment for differences in basicity. Most important, the stabilities of the $M(PNTP)^{2-}$ complexes should reflect the stability of the open species in the equilibrium in Equation 3, thus allowing application of Equations 4 and 5.

Indeed, determination of the stability constants by potentiometric pH titrations (*36*) for a given metal ion from the series Mg^{2+}, Ca^{2+}, Mn^{2+}, Co^{2+}, Ni^{2+}, Cu^{2+}, Zn^{2+}, and Cd^{2+}, resulted in the same stability constant within experimental error for complexes with UTP^{4-}, TTP^{4-} and CTP^{4-} (see column 3 in Table II). There was only a single exception (*36*): $Cu(CTP)^{2-}$ was somewhat more stable than $Cu(UTP)^{2-}$ and $Cu(TTP)^{2-}$.

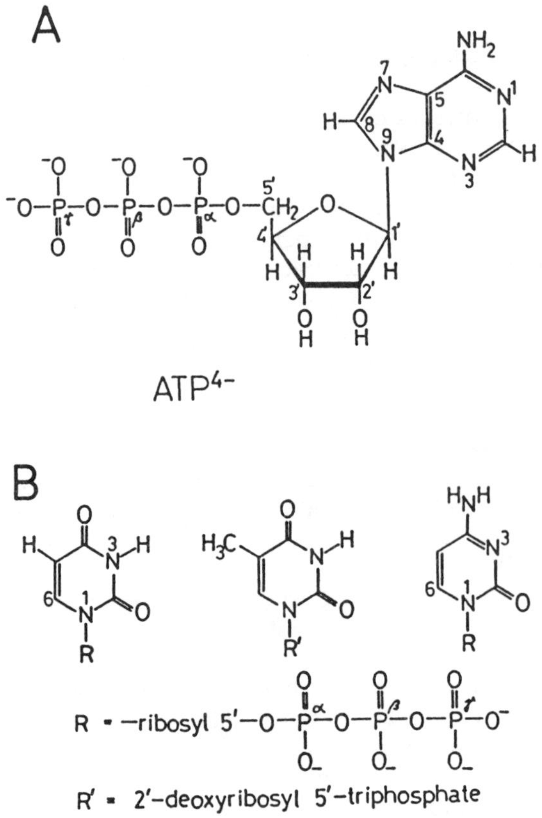

Figure 2. (A) Chemical structure of adenosine 5'-triphosphate (ATP^4-) in the dominating anti conformation (27, 38, 39). (B) Structures of the nucleic base residues of the pyrimidine nucleoside 5'-triphosphates (PNTP^4-).

Table II. Comparison of the Stability of M(ATP)$^{2-}$ Complexes[a] with the Stability of M(PNTP)$^{2-}$ Complexes[a,b] Having only M^{2+}/Phosphate Coordination, and Extent of the Intramolecular Macrochelate Formation in M(ATP)$^{2-}$ Complexes in Aqueous Solution (I = 0.1 M, NaNO$_3$ or NaClO$_4$; 25 °C)

M^{2+}	$\log K^M_{M(ATP)}$ (eq 8a)	$\log K^M_{M(PNTP)}$ $= \log K^M_{M(ATP)_{op}}$[b]	$\log \Delta$ (eq 5)	K_I (eq 4)	% M(ATP)$^{2-}_{cl}$ (eq 3)[c]
Mg^{2+}	4.29±0.02	4.24±0.02	0.05±0.03	0.12±0.07	11±6/13± 6[d]
Ca^{2+}	3.91±0.02	3.90±0.02	0.01±0.03	0.02±0.07	2±6
Mn^{2+}	5.01±0.05	4.93±0.02	0.08±0.06	0.20±0.15	17±10
Co^{2+}	4.97±0.06	4.76±0.02	0.21±0.06	0.62±0.24	38±9/35±10[e]
Ni^{2+}	4.86±0.03	4.50±0.02	0.36±0.04	1.29±0.19	56±4/58[f]
Cu^{2+}	6.34±0.02	5.86±0.02	0.48±0.03	2.02±0.20	67±2/68±4[g]
Zn^{2+}	5.16±0.04	5.02±0.01	0.14±0.04	0.38±0.13	28±7/26±5[e]
Cd^{2+}	5.34±0.02	5.07±0.02	0.27±0.03	0.86±0.12	46±4/50±6[e]; 52[h]

[a]Data are taken from Table IV of ref 36. The errors given are 2 times the standard error of the mean value (2σ) or the sum of the probable systematic errors, whichever is larger. The error limits for log Δ and those of the data in the following columns were calculated according to the error propagation after Gauss. Connected acidity constants are $pK^H_{H_2(ATP)}$ = 4.00 ± 0.01, $pK^H_{H_2(CTP)}$ = 4.55 ± 0.03, and $pK^H_{H(NTP)}$ = 6.50 ± 0.05; for details see ref 36.

[b]Average of the stability constants determined for M(UTP)$^{2-}$, M(TTP)$^{2-}$, and M(CTP)$^{2-}$ (= M(PNTP)$^{2-}$; PNTP^{4-} = pyrimidine-nucleoside 5'-triphosphate), with the exception of Cu(PNTP)$^{2-}$, where only the average of the constants for Cu(UTP)$^{2-}$ and Cu(TTP)$^{2-}$ was used (see text, and ref 36).

[c]Percentage closed species = 100•K_I/(1+K_I). The values before the slash are calculated from the constants given in this table; the values after the slash are from different sources.[d-h]

[d]Calculated from stability constants determined by Frey and Stuehr (40); I = 0.1 M, 15–25 °C; see Table VI in ref 36 or Table 4 in ref 35; error limit: 1σ.

[e]From ref 41. Calculated via the stability of mixed ligand M(ATP)(imidazole)$^{2-}$ and M(UTP)(imidazole)$^{2-}$ complexes. I = 0.1 M, NaNO$_3$; 25 °C; error limits: 3σ.

[f]I = 0.1 M, KNO$_3$; 15 °C; otherwise footnote 'd' applies.

[g]From ref 34; I = 0.1 M, NaNO$_3$; 25 °C.

[h]From ref 42; I = 0.1 M, NaNO$_3$; 25 °C.

Consequently, it was concluded that $32 \pm 6\%$ of $Cu(CTP)^{2-}$ exists in the macrochelated form (Equation 3), with Cu^{2+} not only coordinated to the phosphate chain but also to N-3 of the cytosine residue. Hence, in the macrochelated $Cu(CTP)^{2-}$ species the nucleotide must be present in the less favored *syn* conformation. The energy barrier between the *anti* and *syn* conformations of CTP^{4-} was estimated (36) to about 6 kJ/mol.

Isomeric Equilibria in M(ATP)²⁻ Complexes

This topic has recently been reviewed (35); therefore only additional points together with some prominent features of general importance needed to obtain a coherent view in connection with the following sections are summarized here.

In Table II are listed the stability constants measured for the $M(ATP)^{2-}$ and the $M(PNTP)^{2-}$ complexes (Figure 2) (36). It is evident that in several instances the stability of $M(ATP)^{2-}$ is larger than of $M(PNTP)^{2-}$ leading to values for $\log \Delta > 0$, confirming thus the expectations indicated in the preceding section. These $\log \Delta$ values may be used to calculate with Equation 4 the intramolecular equilibrium constant K_I and the percentage of the closed species occurring in the equilibrium shown in Equation 3; these data are also given in Table II.

It should be emphasized that the results of Table II are based on a large number of experimental data which had been collected with the aim of obtaining values for $\% M(ATP)_{cl}^{2-}$ that are as reliable as possible (36). Indeed, the values given to the right of the slash in the last column of Table II confirm the reliability of the results; these data have independently been determined by other groups or by different methods (34, 40–42), but also *via* potentiometric pH titrations. These facts are important with regard to the following discussion.

Determination of the percentages of $M(ATP)_{cl}^{2-}$ by UV-difference spectrophotometry (7, 43) and ^1H NMR shift experiments (24) gave smaller values for $\% M(ATP)_{cl}^{2-}$ than those listed in Table II. This apparent discrepancy is especially clear-cut for $Mg(ATP)^{2-}$ and $Ni(ATP)^{2-}$ where approximately 0 and 30%, respectively, for the closed species were deduced. Taking into account that ultraviolet absorption and nuclear magnetic resonance spectroscopy techniques, which detect perturbations in the adenine ring, are sensitive mainly to inner sphere coordination of the ring by a metal ion, this observation has led to the suggestion (36) that in addition outer sphere macrochelates are formed. It is evident that the stability increase detected by potentiometric pH titrations encompasses both inner sphere and outer sphere coordination of N-7, i.e. $M(ATP)_{cl/tot}^{2-}$ (= last column in Table II); hence, the difference between $\% M(ATP)_{cl/tot}^{2-}$ and the percentage determined by the spectroscopic methods sensitive mainly to inner sphere coordination, i.e. to $M(ATP)_{cl/i}^{2-}$, should provide the percentage of the N-7 outer sphere coordinated species, $M(ATP)_{cl/o}^{2-}$.

The second column in Table III lists the "best" value presently available (*36*) for % $M(ATP)^{2-}_{cl/i}$. *Via* the procedure detailed in the previous paragraph one can derive % $M(ATP)^{2-}_{cl/o}$ (third column). The difference between 100% and the sum of the percentages of the two macrochelated species gives, then, the percentage of the open complex, $M(ATP)_{op}^{2-}$ (fourth column). Tentative and simplified structures of the two types of macrochelates (*35*) are depicted in Figure 3. It may be mentioned in this connection that very recently evidence has been presented (*44*) that in dilute neutral D_2O solutions *cis*-$Pt(ND_2CH_3)_2^{2+}$ coordinates to purine-nucleoside 5'-triphosphates *via* N-7 and the γ-phosphate group, forming a macrochelate. A further interesting result is the evidence (*45*) that in $Mg(ATP)^{2-}$ phosphate binding occurs as a mixture of β,γ-bidentate and α,β,γ-tridentate complexation. In addition, it was shown (*46*) for Cu^{2+} that in strongly alkaline media the hydroxy groups of the ribose residue become important binding sites.

It is evident that at present only indirect evidence for the formation of N-7 outer sphere macrochelates (Figure 3B) can be presented. It should be emphasized, though, that the differences between $M(ATP)^{2-}_{cl/tot}$ and $M(ATP)^{2-}_{cl/i}$ are certainly beyond the experimental errors, e.g., for the complexes (*35, 36*) with Ni^{2+} and Mg^{2+}. Hence, it is clear that at least with some metal ions this species is formed, though the numbers in Table III may change somewhat with the determination of more precise values for % $M(ATP)^{2-}_{cl/i}$. The possible biological significance of this species has been discussed (e.g., with regard to kinases; *35*). It is interesting to note that the sequence of total (last column in Table II) and inner sphere closed forms (second column in Table III) follows the usual stability series for dipositive first-row transition metal ions (*47–49*), including the stability constants for imidazole binding (*50–52*) and the relative placements of Zn^{2+} and Cd^{2+}. It may be added that this solution information has been rationally exploited (*53*) in the preparation of novel mixed-metal ion–ATP complexes.

At this point it should be emphasized that the differences in free energy (ΔG^0) connected with the formation of low percentages of a macrochelate, $M(NP)_{cl}$, are small: e.g., 5, 21 and 50% $M(NP)_{cl}$ correspond to changes in ΔG^0 of only −0.1, −0.6 and −1.7 kJ/mol, respectively (see Table IV). It is evident that for an enzymic reaction a relatively small amount (say, e.g., 20%) of a given isomer of a complex occurring in fast equilibrium is enough to serve as substrate. Hence, the energies involved between the formation of different structural M(NP) isomers may be low, and this appears to be ideal to promote enzymic selectivity. Connecting a series of such equilibria would result in a rather high selectivity, despite the small differences in free energy (ΔG^0). Clearly, high degrees of formation of a macrochelate are connected with substantial changes in complex stability (*37*): e.g., 90% $M(NP)_{cl}$ corresponds to log $\Delta = 1$ and $\Delta G^0 = -5.7$ kJ/mol; 99% $M(NP)_{cl}$ requires log $\Delta = 2$ and $\Delta G^0 = -11.4$ kJ/mol (Table IV).

Table III. Estimates for the Degree of Formation of N-7 Inner Sphere, $M(ATP)^{2-}_{cl/i}$, and Outer Sphere Macrochelates, $M(ATP)^{2-}_{cl/o}$, in $M(ATP)^{2-}$ Systems in Aqueous Solution (~25°C; I ~ 0.1 M); % $M(ATP)^{2-}_{op}$ for the 'open' Species (Equation 3) is Given for Completeness[a]

M^{2+}	% $M(ATP)^{2-}_{cl/i}$	% $M(ATP)^{2-}_{cl/o}$	% $M(ATP)^{2-}_{op}$
Mg^{2+}	0	10	90
Ca^{2+}	0	~0	~100
Mn^{2+}	~10	~10	80
Co^{2+}	~25	~15	60
Ni^{2+}	30	25	45
Cu^{2+}	67	~0	33
Zn^{2+}	15	15	70
Cd^{2+}	30	20	50

[a]Abstracted from Table VI of (36); for further details see (35, 36).

Table IV. Relation Between log Δ (Equation 5), K_I (Equation 4), the Extent of Macrochelate Formation, $M(NP)_{cl}$ (Equation 3), and ΔG^0 (25 °C)

log Δ	K_I	% $M(NP)_{cl}$	ΔG^0 (kJ/mol)
0.02	0.05	5	−0.1
0.05	0.12	11	−0.3
0.1	0.26	21	−0.6
0.2	0.58	37	−1.1
0.3	1.0	50	−1.7
0.6	3.0	75	−3.4
1.0	9.0	90	−5.7
2.0	99	99	−11.4

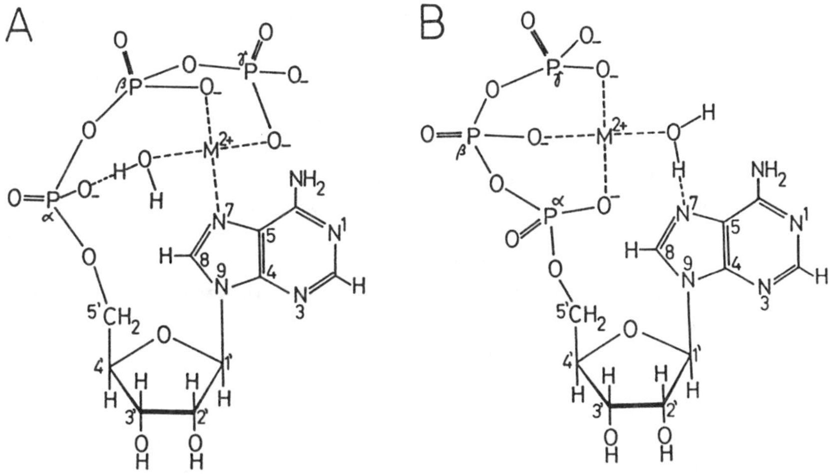

Figure 3. Tentative and simplified structures for the macrochelated inner sphere (A) and outer sphere (B) M(ATP)$^{2-}$ species (35, 36). It should be noted that the terms inner sphere and outer sphere are used here with regard to M^{2+}/N-7 coordination. The depicted M^{2+}/triphosphate coordinations follow the suggestions of Martin and Mariam (38): If an intramolecular direct M^{2+}/N-7 coordination occurs, then it is sterically more favorable to have a water molecule between M^{2+} and the α-phosphate group as shown in A. With outer sphere N-7 binding, as shown in B, inner sphere coordination of all three phosphate groups is suggested (38). However, other isomers differing in the phosphate coordination are possible; e.g. direct β,γ-phosphate and N-7 coordination leaving the α group free. Finally it should be noted that for the sake of clarity in the above structures the equatorial positions of an octahedral coordination sphere are used, but binding to other positions is of course also possible (giving rise then to further isomers). Reproduced from Eur. J. Biochem. (35) with permission.

That N-7 coordination is the "weak" point in macrochelate formation of the M(ATP)$^{2-}$ complexes, in accordance with the results of Tables II and IV, may be demonstrated by the release of N-7 upon imidazole (Im) coordination (41). The results for the Cd(ATP)$^{2-}$/Im system are summarized in Figure 4. In the upper part of the figure the chemical shifts of H-8, H-2 and H-1' are plotted, and in the lower part the degree of formation of Cd(ATP)(Im)$^{2-}$ is shown. The dependence of each on the concentration of imidazole is plotted. At [Im] = 0 the addition of one equivalent of Cd^{2+} to ATP^{4-} leads to Cd(ATP)$^{2-}$. Due to macrochelate formation with N-7 (see Tables II and III), the expected downfield shift of H-8 is also observed. Increasing amounts of imidazole lead to an increasing degree of formation of Cd(ATP)(Im)$^{2-}$. The corresponding upfield shift of H-8 demonstrates the release of N-7 from the coordination sphere of the metal ion in this mixed-ligand complex. Consideration of finer details in Figure 4 indicates that Cd(ATP)$^{2-}$ is partly self-associated under the experimental conditions, and also that in Cd(ATP)(Im)$^{2-}$ some intramolecular stacking between the aromatic ring systems of the two ligands occurs. These and other aspects of this system have been discussed (41).

Similar ^{1}H NMR experiments reveal that imidazole releases inner sphere N-7 in Zn(ATP)$^{2-}$, and ammonia releases N-7 in Cd(ATP)$^{2-}$, upon formation of Zn(ATP)(Im)$^{2-}$ and Cd(ATP)(NH$_3$)$^{2-}$, respectively (41). Similar observations were made with OH$^-$ as second ligand (42). Stability constant comparisons (41) reveal that addition of imidazole at least weakens and possibly also eliminates N-7 backbonding in Ni(ATP)$^{2-}$ and Cu(ATP)$^{2-}$ (see also 54). To conclude (41), the above results as well as other previous studies (55–59) indicate that mixed-ligand complex formation between M(ATP)$^{2-}$ and a monodentate or bidentate ligand leads in general to a release of the adenine moiety from the coordination sphere of the metal ion. This release of N-7 is apparently rather independent of the size and properties of the second ligand; it has now been observed for ligands as different as OH$^-$, NH$_3$, imidazole, 2,2'-bipyridyl, 1,10-phenanthroline, and tryptophanate. These results suggest that when bound to an enzyme the binary M(ATP)$^{2-}$ unit may exist as a closed N-7 macrochelate only if no enzyme groups coordinate directly to the metal ion.

Relationship Between Complex Stability and Ligand Basicity for Monophosphate Ligands. Pyrimidine-Nucleoside 5'-Monophosphates are Simple Phosphate-Coordinators

In the case of nucleoside triphosphates the stability constants of different M(NTP)$^{2-}$ complexes for a given metal ion may be directly compared, because the basicity of the terminal γ-phosphate group in the H(NTP)$^{3-}$ species is very similar (pK$^H_{H(NTP)}$ = 6.50 ± 0.05). With nucleoside monophosphates (NMP^{2-}) the situation is more complicated because the basicity of the phosphate group varies due to its closer proximity to the

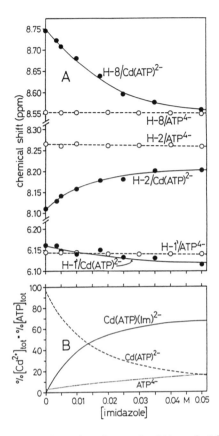

Figure 4. Comparison of (A) the chemical shifts of Cd(ATP)²⁻ under the influence of increasing concentrations of imidazole, with (B) the resulting increasing concentration of the ternary Cd(ATP)(Im)²⁻ complex in D₂O at pD 8.4. (A) Dependence of the chemical shifts of H-2, H-8, and H-1' of Cd(ATP)²⁻ (●) ([Cd²⁺]ₜₒₜ = [ATP]ₜₒₜ = 5 × 10⁻³ M; degree of formation of Cd(ATP)²⁻ about 96%, see below in (B)) on increasing concentrations of imidazole in D₂O at pD 8.4 (I = 0.1 M, NaNO₃; 27 °C). The three nearly horizontal, broken lines represent the chemical shifts of H-2, H-8, and H-1' of uncomplexed ATP (5 × 10⁻³ M) also at pD 8.4 and under the influence of increasing concentrations of imidazole (O) (41). (B) Effect of increasing concentration of imidazole (pD 8.4; I = 0.1 M; 25 °C) on the concentration of the species present in a D₂O solution of Cd²⁺ and ATP (each 5 × 10⁻³ M). The results were computed with the equilibrium constants given in (41), and they are presented as the percentage of the total Cd²⁺ present (= total ATP). The slight self-association (24) is neglected [about 85% of Cd(ATP)²⁻ is in the monomeric form (41)], and it is assumed in the calculations that the change from H₂O to D₂O does not influence the sizes of the stability constants of the complexes [the influence on the acidity constants is considered (41)]. Reproduced from (41) of the American Chemical Society.

organic part of the ligand: $pK_{H(NMP)}^{H} \approx 5.7$ to 6.5 (27, 60). Consequently, application of Equation 4 to determine the position of the intramolecular equilibrium shown in Equation 3 is only possible if log Δ is calculated from 'basicity-adjusted' stability constants. This means that it is necessary to establish for monophosphate esters (R-MP^{2-}) the relationship between complex stability and phosphate group basicity.

To achieve this goal the acidity constants, $K_{H(R-MP)}^{H}$, and the corresponding stability constants, $K_{M(R-MP)}^{M}$, of several metal ion complexes of 4-nitrophenyl phosphate (NPheP^{2-}), phenyl phosphate (PheP^{2-}), n-butyl phosphate (BuP^{2-}) and D-ribose 5'-monophosphate (RibMP^{2-}) were measured by potentiometric pH titrations in aqueous solutions (I = 0.1 M, NaNO$_3$; 25 °C; 60). As expected for a family of related ligands (37) plots of log $K_{M(R-MP)}^{M}$ versus $pK_{H(R-MP)}^{H}$ result in straight lines (60). The available equilibrium data for the corresponding systems with methyl phosphate (CH$_3$OPO$_3$$^{2-}$; 61, 62) and hydrogen phosphate (HPO$_4$$^{2-}$; 13) also fit these straight lines (60). This proves that we are indeed dealing with simple monophosphate complexes, the stability of which is not influenced by steric restrictions or other additional interactions, but is solely determined by the basicity of the phosphate group in the R-MP^{2-} ligand.

For the pyrimidine nucleotides, uridine 5'-monophosphate (UMP^{2-}) and thymidine 5'-monophosphate (TMP^{2-}), no metal ion-base interaction is expected in M(R-MP) complexes (see the nucleic base structures in Figure 2) because the nucleic base residue becomes an effective ligand only after deprotonation of the H^{+}(N-3) unit (24, 63). Indeed, this is confirmed: in plots of log $K_{M(R-MP)}^{M}$ versus $pK_{H(R-MP)}^{H}$ the data fit within experimental error to the above-mentioned straight lines (60) for complexes with Mg^{2+}, Ca^{2+}, Sr^{2+}, Ba^{2+}, Mn^{2+}, Co^{2+}, Ni^{2+}, Cu^{2+}, Zn^{2+}, and Cd^{2+}, consistent with the expectation that UMP^{2-} and TMP^{2-} show the properties of simple monophosphate esters in M(R-MP) complexes.

In summary, the best straight lines were calculated (60) for the log $K_{M(R-MP)}^{M}$ versus $pK_{H(R-MP)}^{H}$ plots and the above-mentioned metal ions and their complexes with the following six ligands: NPheP^{2-}, PheP^{2-}, BuP^{2-}, RibMP^{2-}, UMP^{2-}, and TMP^{2-} (e.g. Figure 7; vide infra). The equations for the resulting straight lines are listed in Table V (60, 64). These data allow calculation of the stability of an M(R-MP) complex with coordination only by phosphate, in the pH range 5 to 7, using the known pK$_a$ value of an H(R-MP)$^-$ species.

For example, application of these results to the complexes formed with cytidine 5'-monophosphate (CMP^{2-}) allows the demonstration that the cytosine residue (for its structure, see Figure 2) does not participate in complex formation in the M(CMP) complexes considered, as there is no sign of a stability increase (60). The stability constants, determined by potentiometric pH titrations for the equilibrium shown in Equation 9, are within experimental error identical with the stability constants calculated

$$M^{2+} + NMP^{2-} \rightleftarrows M(NMP) \qquad (9a)$$

$$K^M_{M(NMP)} = \frac{[M(NMP)]}{([M^{2+}][NMP^{2-}])} \tag{9b}$$

with the pK_a value due to the phosphate group of $H(CMP)^-$ and the straight-line equations given in Table V. This means that log Δ according to Equation 5 is zero within experimental error (see Table VI). CMP^{2-} therefore displays the properties of a simple phosphate monoester (60) toward the metal ions considered. Hence, the pyrimidine-like N-3 of the cytosine residue does not coordinate in the M(CMP) species. This result agrees with the structures of pyrimidine nucleotides in solution: they occur predominantly in the *anti* conformation (38), i.e. N-3 is pointing away from the phosphate group. However, this does not mean that N-3 has no affinity for metal ions. From spectrophotometric measurements (43) with Ni^{2+} and ^1H NMR shift experiments (24) with Cd^{2+} it is known that cytidine (Cyd) complexes may form, and the stability constant of $Cu(Cyd)^{2+}$ has been determined (65). This observation for M(CMP) complexes clearly shows that the presence of additional binding sites is not enough for the formation of macrochelates. A conformation of the nucleotide allowing a simultaneous interaction at different sites is also necessary. In fact, the energy barrier between the *anti* and *syn* conformations is considerable (about 6 kJ/mol for CTP^{4-}; 36).

Some of the structures which appear in the literature for simple phosphate monoester complexes are summarized in Figure 5. Based on the comparison of the slopes of log K^M_{ML} versus pK^H_{HL} plots it is tentatively suggested (60) that structure A of Figure 5 is of little relevance in aqueous solution. (However, this does not mean that such a four-membered ring may not be formed as an intermediate in low concentration during a reaction.) Instead, structures B and C are the important binding modes of monophosphate groups to substitution-labile divalent metal ions. In addition, there are indications (61, 62, 66) that various amounts of outer sphere species may exist in equilibrium. This agrees with the structure of barium adenosine 5'-monophosphate heptahydrate in the solid state, in which the completely hydrated Ba^{2+} interacts only in an outer sphere manner with the nucleotide, i.e. in particular *via* phosphate-hydrogen bonds of coordinated water molecules (67). The concentration of such outer sphere species seems to be appreciable with Ca^{2+}, Mg^{2+}, Mn^{2+}, Co^{2+} (61) and Ni^{2+} (61, 62), but negligible with Cu^{2+} and Zn^{2+} (61). It is further concluded (62) that the higher the charge of a phosphate, the more predominant are inner sphere complexes; e.g., Ni^{2+} coordinates to the phosphate chain of ATP^{4-} overwhelmingly in an inner sphere fashion (38, 62).

From the discussion in the previous paragraph it follows that it is necessary to assume that M(R-MP) complexes in aqueous solution participate in intramolecular equilibria between several phosphate-metal ion binding modes. The ratios of these different binding modes for M(R-MP) complexes are expected to depend on the metal ion involved, but not on the nature of the noncoordinating residue R. In connection with the

Table V. Correlation of Metal Ion-Phosphate Coordination and Phosphate Group Basicity of Monophosphate Monoesters

M^{2+}	m	b	SD
Mg^{2+}	0.224 ± 0.027	0.174 ± 0.167	0.014
Ca^{2+}	0.156 ± 0.039	0.487 ± 0.239	0.018
Sr^{2+}	0.089 ± 0.034	0.691 ± 0.206	0.016
Ba^{2+}	0.073 ± 0.036	0.706 ± 0.217	0.017
Mn^{2+}	0.250 ± 0.048	0.607 ± 0.293	0.022
Co^{2+}	0.230 ± 0.057	0.510 ± 0.345	0.024
Ni^{2+}	0.282 ± 0.045	0.201 ± 0.275	0.021
Cu^{2+}	0.453 ± 0.056	0.055 ± 0.340	0.026
Zn^{2+}	0.321 ± 0.057	0.125 ± 0.345	0.027
Cd^{2+}	0.317 ± 0.042	0.467 ± 0.253	0.019

The slopes (m) and intercepts (b) for the straight baselines (log $K^M_{M(R-MP)}$ versus $pK^H_{H(R-MP)}$) are calculated from the equilibrium constants determined for the simple phosphate monoesters 4-nitrophenyl phosphate, phenyl phosphate, n-butyl phosphate, D-ribose 5'-monophosphate, uridine 5'-monophosphate, and thymidine 5'-monophosphate (I = 0.1 M, $NaNO_3$; 25 °C; see 60). The column at the right lists the standard deviations (SD) resulting from the differences between the experimental and calculated values for these six ligand systems.

The data are abstracted from Tables V and VI in (60). Straight-line equation: $y = m \cdot x + b$, where x represents the pK_a value of any phosphate monoester; the errors given for m and b correspond to one standard deviation (1σ). The listed SD values times 2 or 3 are considered as reasonable error limits for any stability constant calculation in the pK_a range 5 to 7.

Reprinted from (64) of the American Chemical Society.

Table VI. Logarithms of the Stability Constants of M(CMP) Complexes (Equation 9) as Determined by Potentiometric pH Titrations (Exptl) in Water (25 °C; I = 0.1 M, NaNO$_3$)[a]

M^{2+}	log K$^M_{M(CMP)}$		log Δ =
	Exptl[a]	Calcd[b]	log K$_{Exptl}$ − log K$_{Calcd}$[c]
Mg^{2+}	1.54 ± 0.05	1.55(1.56) ± 0.04	−0.01 ± 0.06
Ca^{2+}	1.40 ± 0.05	1.45(1.45) ± 0.05	−0.05 ± 0.07
Sr^{2+}	1.17 ± 0.04	1.24(1.24) ± 0.05	−0.07 ± 0.06
Ba^{2+}	1.11 ± 0.03	1.15(1.16) ± 0.05	−0.04 ± 0.06
Mn^{2+}	2.10 ± 0.04	2.14(2.15) ± 0.07	−0.04 ± 0.08
Co^{2+}	1.86 ± 0.05	1.92(1.93) ± 0.07	−0.06 ± 0.09
Ni^{2+}	1.94 ± 0.06	1.94(1.95) ± 0.06	0.00 ± 0.08
Cu^{2+}	2.84 ± 0.06	2.84(2.86) ± 0.08	0.00 ± 0.10
Zn^{2+}	2.06 ± 0.05	2.10(2.11) ± 0.08	−0.04 ± 0.09
Cd^{2+}	2.40 ± 0.08	2.42(2.43) ± 0.06	−0.02 ± 0.10

For comparison the calculated stability constants (Calcd)[b] based on the basicity of the phosphate group and the baseline equations of Table V are also given (see column 3).
[a]The errors given are 3 times the standard error of the mean or the sum of the probable systematic errors, whichever is larger. The macroacidity constants of H$_2$(CMP)$^\pm$ are: pK$^H_{H(CMP)}$ = 6.19 ± 0.02 and pK$^H_{H_2(CMP)}$ = 4.33 ± 0.04.
[b]The first value is calculated with the microconstant pk$^{CMP}_{CMP \cdot H}$ = 6.15 (= pK$^H_{H(UMP)}$), i.e. with the acidity constant corrected for the influence of the protonated base residue (60). The values in parentheses are based on pK$^H_{H(CMP)}$ = 6.19. The error limits correspond to 3 times the standard deviations listed in Table V.
[c]The error limits for these differences were calculated according to the error propagation after Gauss.
Reprinted from (60) of the American Chemical Society.

application of the straight-line equations (Table V) to nucleoside monophosphate complexes, the following consideration is important: in an M(NMP) complex in the open form (Equation 3) the metal ion will coordinate to the phosphate group in the same way as with simple phosphate monoesters. Should macrochelate formation in backbound M(NMP) species alter the phosphate inner sphere : outer sphere ratio to some extent, then the percentage of the total closed form calculated for the corresponding M(NMP) complex (see the following sections) would still be correct. This is because any alteration of the phosphate inner sphere : outer sphere ratio would occur at the cost of the stability increase, i.e. log Δ (Equation 5), on which the calculation of the extent of macrochelation is based (Equation 4).

Definite Evidence for Metal Ion-Base-Backbinding to N-7, and the Extent of Macrochelate Formation in M(AMP) Species

Replacement of N-7 in adenosine 5'-monophosphate (AMP^{2-}) by a CH unit gives tubercidin 5'-monophosphate ($TuMP^{2-}$ = 7-deaza-AMP^{2-}) (Figure 6). Comparison of the corresponding M^{2+} complexes should allow an evaluation of the influence of N-7 on the coordinating properties of AMP^{2-}. Moreover, if N-7 is responsible for base-backbinding of the phosphate-coordinated metal ion in certain M(AMP) species, then $TuMP^{2-}$ should behave as a simple phosphate monoester ligand. This notion is indeed correct. In the experimental results (64) plotted in Figure 7, it is evident that the points due to the $TuMP^{2-}$ systems lie on the reference lines.

A careful evaluation (64) of the stability constant data for all the metal ion–$TuMP^{2-}$ complexes studied is contained in Table VII. The stability constants determined by potentiometric pH titration (I = 0.1 M, $NaNO_3$; 25 °C) for log $K^M_{M(TuMP)}$ (Equation 9) are, within experimental error, identical with those calculated via the pK_a value of the monoprotonated phosphate group of TuMP and the straight-line equations of Table V; i.e., log Δ (Equation 5) is zero within the error limits. Consequently, there is no further interaction of any kind in these M(TuMP) complexes besides the expected phosphate-metal ion binding. This conclusion also holds for N-1. Although one might be tempted to assume that in M(TuMP) the nucleotide could be forced into a *syn* conformation to allow backbinding to the rather basic N-1 ($pK^{TuMP\cdot H}_{H\cdot TuMP\cdot H}$ = 5.36; 64), this clearly does not occur.

However, a metal ion coordinated to the phosphate group of AMP^{2-}, in the preferred *anti* conformation, may reach N-7 of the base residue. In those cases where this interaction occurs, an increase in complex stability must result (37). A clear example is seen in Figure 7, where the point for the Cu^{2+}/AMP system is significantly above the reference line. This demonstrates that Cu(AMP) is more stable than would be expected considering only the basicity of the phosphate group ($pK^H_{H(AMP)}$ = 6.21; 64). In fact, the vertical distance between this data point and the reference line

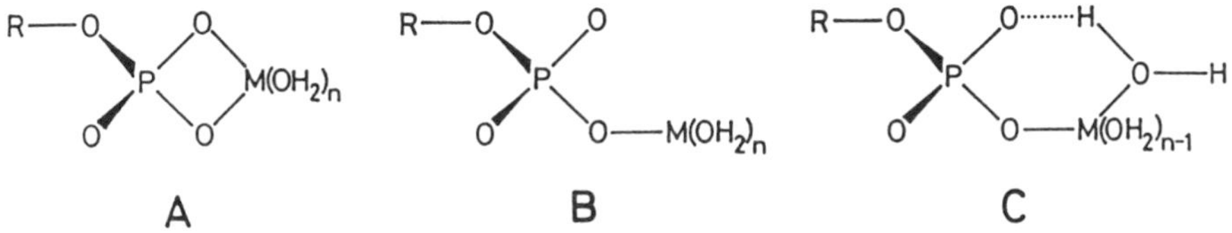

Figure 5. Metal ion binding modes in M(R-MP) complexes. Structure A is considered of little importance in aqueous solution, while structures B and C are proposed as the important binding modes (see text). In addition, there are indications (61, 62, 66) that in equilibrium an outer sphere binding mode also may occur; i.e., the phosphate oxygens and the metal ion are separated by water molecules. Reproduced from (60) of the American Chemical Society.

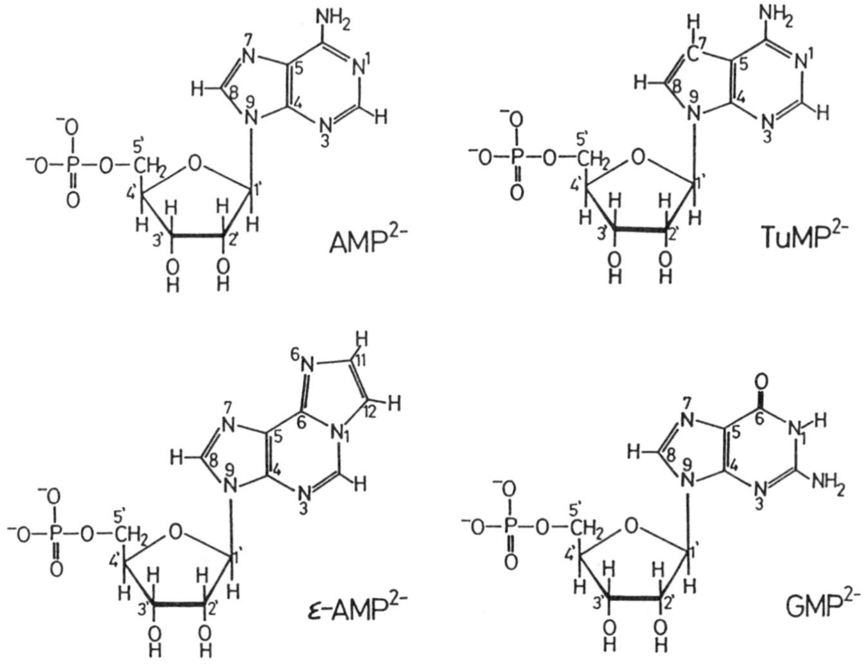

Figure 6. Chemical structures of adenosine 5'-monophosphate (AMP^{2-}), tubercidin 5'-monophosphate ($TuMP^{2-}$ = 7-deaza-AMP^{2-}), 1,N^6-ethenoadenosine 5'-monophosphate (ε-AMP^{2-} = 1,N^6-etheno-AMP^{2-}), and guanosine 5'-monophosphate (GMP^{2-}) in their dominating anti conformations (27, 38).

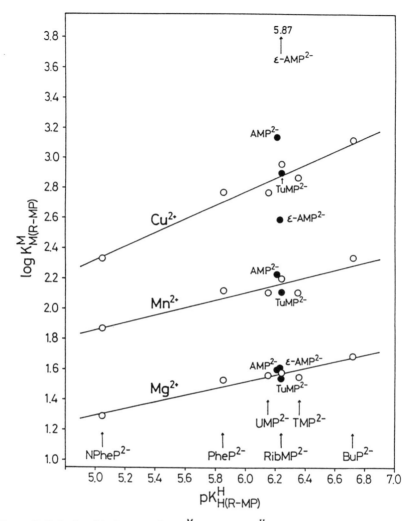

Figure 7. *Relationship between* log $K_{M(R-MP)}^M$ *and* p$K_{H(R-MP)}^H$ *for the 1:1 complexes of* Mg^{2+}, Mn^{2+}, *and* Cu^{2+} *with some simple phosphate monoester ligands (R-*MP^{2-}*): 4-nitrophenyl phosphate (NPheP^{2-}), phenyl phosphate (PheP^{2-}), uridine 5'-monophosphate (UMP^{2-}), D-ribose 5'-monophosphate (RibMP^{2-}), thymidine 5'-monophosphate (TMP^{2-}), and n-butyl phosphate (BuP^{2-}) (from left to right) (O). Least-squares lines are drawn through the corresponding data sets, which are taken from (60); the equations for these base lines are given in Table V. The points due to the complexes formed with TuMP^{2-}, AMP^{2-}, and ε-AMP^{2-} (●) are inserted for comparison; the stability constants for the M(TuMP) complexes are plotted versus the microconstant* p$k_{TuMP \cdot H}^{TuMP}$ = 6.24 *(see 64); the other equilibrium constants for these three NMPs are from Tables VII–IX. All plotted equilibrium constant values refer to aqueous solutions at 25 °C and I = 0.1 M (NaNO$_3$). Reproduced from (64) of the American Chemical Society.*

Table VII. Stability Constant Comparisons for the M(TuMP)
Complexes (Equation 9) Between the Measured Stability Constants
(Exptl)[a] and the Calculated Stability Constants (Calcd) Based on the
Basicity of the Phosphate Group[b] and the Baseline Equations
of Table V (25 °C; I = 0.1 M, NaNO₃)

	log $K^M_{M(TuMP)}$		log Δ =
M^{2+}	Exptl[a]	Calcd[b]	log K_{Exptl} − log K_{Calcd}[c]
Mg^{2+}	1.54 ± 0.06	1.57 ± 0.04	-0.03 ± 0.07
Ca^{2+}	1.43 ± 0.06	1.46 ± 0.05	-0.03 ± 0.08
Sr^{2+}	1.24 ± 0.04	1.25 ± 0.05	-0.01 ± 0.06
Ba^{2+}	1.13 ± 0.06	1.16 ± 0.05	-0.03 ± 0.08
Mn^{2+}	2.11 ± 0.05	2.17 ± 0.07	-0.06 ± 0.09
Co^{2+}	1.94 ± 0.04	1.95 ± 0.07	-0.01 ± 0.08
Ni^{2+}	2.04 ± 0.08	1.96 ± 0.06	0.08 ± 0.10
Cu^{2+}	2.90 ± 0.08	2.88 ± 0.08	0.02 ± 0.11
Zn^{2+}	2.11 ± 0.05	2.13 ± 0.08	-0.02 ± 0.09
Cd^{2+}	2.42 ± 0.07	2.45 ± 0.06	-0.03 ± 0.09

[a]The errors given are 3 times the standard error of the mean or the sum of the probable systematic errors, whichever is larger. The macroacidity constants of $H_2(TuMP)^{\pm}$ are $pK^H_{H_2(TuMP)} = 5.28 \pm 0.02$ and $pK^H_{H(TuMP)} = 6.32 \pm 0.01$ (64).
[b]Calculated with the microconstant $pk^{TuMP}_{TuMP \cdot H} = 6.24$ (= $pK^H_{H(RibMP)}$) (64); the error ranges are 3 times the SD values listed in the column at the right in Table V.
[c]The error limits for these differences were calculated according to the error propagation after Gauss.

Reprinted from (64) of the American Chemical Society.

corresponds to log Δ (Equation 5), which allows calculation of K_I (Equation 4) and thus the percentage of the macrochelated form in the equilibrium shown in Equation 3.

The results for the ten metal ion/AMP systems studied (64) are summarized in Table VIII. The M(AMP) complexes of the alkaline earth ions have log Δ values which are zero within the error limits, showing that the open, phosphate-coordinated species strongly dominate and that at most traces of base-backbound isomers can occur. With the divalent 3d transition metal ions and Zn^{2+} or Cd^{2+}, appreciable amounts of macrochelated species form. In the case of Mn(AMP) the result is just at the limit of significance, but a certain degree of formation of N-7 backbound species is most likely (for the arguments see 64). That the extent of base-backbinding is larger with Ni(AMP) than with Cu(AMP) is probably a result of the different coordination geometries of these two metal ions, and it may be explained by statistical considerations (64). The near identity of the stability enhancements (log Δ) for the complexes with Co^{2+}, Zn^{2+}, and Cd^{2+} (Table VIII) is a reflection of the similar affinity of these ions towards imidazole nitrogen donors (52).

Enhanced Base-Backbinding in Complexes of 1,N⁶-Ethenoadenosine Phosphates Compared with the Parent Nucleotides

That seemingly small structural alterations may affect the metal ion-coordinating properties of nucleotides very strongly is exemplified by the formation of the 1,N⁶-etheno bridge in adenosine nucleotides. Insertion of this bridge into AMP^{2-} leads to the so-called $\varepsilon\text{-}AMP^{2-}$. From the structure shown in Figure 6 it is evident that a 1,10-phenanthroline-like binding site at the nucleic base residue is created in this way. Such 1,N⁶-etheno-adenine derivatives are often employed as fluorescent probes in enzymic reactions which depend on the presence of an adenine cofactor.

The results plotted in Figure 7 show that the phenanthroline-like N-6,N-7 site at the base leads to a very significant increase in stability of the $M(\varepsilon\text{-}AMP)$ species with Mn^{2+} and Cu^{2+}, compared with the parent M(AMP) system. The available stability constants (64, 68–70) have been (in part newly) evaluated. The results regarding the extent of macrochelate formation in several metal ion systems with $\varepsilon\text{-}AMP^{2-}$ and $\varepsilon\text{-}ATP^{4-}$ are summarized in Table IX together with the corresponding data for the parent M(NP) complexes. It is evident that the stability increase (log Δ) in several instances is very large, as is the percentage of the macrochelated form (Equation 3).

The metal ion-coordinating properties of ε-adenine derivatives have recently been reviewed (29). Hence, I emphasize here only that ε-adenine derivatives should never be employed as probes in enzymic systems for the naturally-occurring cofactors in the presence of Zn^{2+} or any transition metal ion, as the structures of the complexes in solution are certainly different. Such studies in the presence of Mg^{2+} or Ca^{2+} appear feasible,

Table VIII. Comparison of the Measured Stability, $K^M_{M(AMP)}$, of the M(AMP) Complexes[a] with the Calculated Stability, $K^M_{M(AMP)_{op}}$, for an Isomer with only M^{2+}/Phosphate Coordination,[b] and Extent of the Intramolecular Macrochelate Formation (Equation 3) in the M(AMP) Complexes at 25 °C and I = 0.1 M (NaNO$_3$)

M^{2+}	$\log K^M_{M(AMP)}$ (eq 8,9)[a]	$\log K^M_{M(AMP)_{op}}$ (eq 7)[b]	$\log \Delta$ (eq 5)[c]	K_I (eq 4)	% M(AMP)$_{cl}$ (eq 3)
Mg^{2+}	1.60 ± 0.02	1.57 ± 0.04	0.03 ± 0.04	0.07 ± 0.10	$0(7{\pm}9/{\leq}15)$
Ca^{2+}	1.46 ± 0.01	1.46 ± 0.05	0.00 ± 0.05	0.00 ± 0.12	$0({\leq}11)$
Sr^{2+}	1.24 ± 0.01	1.24 ± 0.05	0.00 ± 0.05	0.00 ± 0.12	$0({\leq}11)$
Ba^{2+}	1.17 ± 0.02	1.16 ± 0.05	0.01 ± 0.05	0.02 ± 0.13	$0({\leq}13)$
Mn^{2+}	2.23 ± 0.01	2.16 ± 0.07	0.07 ± 0.07^e	0.17 ± 0.19^e	15 ± 14^e
Co^{2+}	2.23 ± 0.02	1.94 ± 0.07	0.29 ± 0.07	0.95 ± 0.33	49 ± 9
Ni^{2+}	2.49 ± 0.02	1.95 ± 0.06	0.54 ± 0.06	2.47 ± 0.50	71 ± 4
Cu^{2+}	3.14 ± 0.01	2.87 ± 0.08	0.27 ± 0.08	0.86 ± 0.35	46 ± 10
Zn^{2+}	2.38 ± 0.07^d	2.12 ± 0.08	0.26 ± 0.11	0.82 ± 0.45	45 ± 13
Cd^{2+}	2.68 ± 0.02	2.44 ± 0.06	0.24 ± 0.06	0.74 ± 0.25	43 ± 8

[a]Determined in aqueous solution by potentiometric pH titrations. The errors given are 3 times the standard error of the mean or the sum of the probable systematic errors, whichever is larger. The acidity constants of $H_2(AMP)^\pm$ are $pK^H_{H_2(AMP)} = 3.84 \pm 0.02$ and $pK^H_{H(AMP)} = 6.21 \pm 0.01$ (64).
[b]Calculated with $pK^H_{H(AMP)} = 6.21$ and the baseline equations of Table V; the error limits correspond to 3 times the SD values given in the column at the right in Table V.
[c]The errors given here and in the other two columns at the right were calculated according to the error propagation after Gauss by using the errors listed in the second and third columns.
[d]Regarding experimental difficulties, see (64). The $\log K^{Zn}_{Zn(AMP)}$ values determined in aqueous solution with NaNO$_3$ and NaClO$_4$ as background electrolyte (I = 0.1 M) were 2.41 $\pm$ 0.10 and 2.34 $\pm$ 0.06, respectively; the value given above is the overall average.
[e]This result is most probably significant; with 2σ as error limits the data are $\log \Delta = 0.07 \pm 0.04$, $K_I = 0.17 \pm 0.11$, and % Mn(AMP)$_{cl}$ = 15 $\pm$ 8 (see also (64)).
Reprinted from (64) of the American Chemical Society.

Table IX. Comparison of the Extent of Macrochelate Formation Between M(NP) Complexes of 1,N⁶-Ethenoadenosine Phosphates and the Parent Nucleotides (25 °C; I = 0.1 M)

M^{2+}	$\log \Delta$ (eq 5)	K_I (eq 4)	% $M(NP)_{cl}$ (eq 3)	$\log \Delta$ (eq 5)	K_I (eq 4)	% $M(NP)_{cl}$ (eq 3)
	$M(\varepsilon\text{-AMP})$[a]			$M(AMP)$[b]		
Mg^{2+}	0.04±0.04	0.10±0.10	9±9(≤17)	0.03±0.04	0.07±0.10	7±9(≤15)
Mn^{2+}	0.43±0.08	1.69±0.50	63±7	0.07±0.07	0.17±0.19	15±14
Co^{2+}	~1.6	~40	~98	0.29±0.07	0.95±0.33	49±9
Ni^{2+}	~2	~100	~99	0.54±0.06	2.47±0.50	71±4
Cu^{2+}	2.99±0.08	976±180	99.9±0.1	0.27±0.08	0.86±0.35	46±10
Zn^{2+}	1.06±0.09	10.5±2.4	91±2	0.26±0.11	0.82±0.45	45±13
	$M(\varepsilon\text{-ATP})^{2-}$ (cf.[c])			$M(ATP)^{2-}$ (cf.[d])		
Mg^{2+}	0.00±0.03	0.00±0.07	0±7	0.05±0.03	0.12±0.07	11±6
Ca^{2+}	~−0.16	0	0	0.01±0.03	0.02±0.07	2±6
Mn^{2+}	0.17±0.10	0.48±0.35	32±16	0.08±0.06	0.20±0.15	17±10
Co^{2+}	~0.34	~1.2	~55	0.21±0.06	0.62±0.24	38±9
Ni^{2+}	~1.3	~19	~95	0.36±0.04	1.29±0.19	56±4
Cu^{2+}	3.1±1	~1300	>99.2	0.48±0.03	2.02±0.20	67±2
Zn^{2+}	0.42±0.09	1.63±0.55	62±8	0.14±0.04	0.38±0.13	28±7

[a]See Figure 6. The data are abstracted from Table VII in (64). The acidity constants of $H_2(\varepsilon\text{-AMP})^{\pm}$ are $pK^H_{H_2(\varepsilon\text{-AMP})} = 4.23 \pm 0.02$ and $pK^H_{H(\varepsilon\text{-AMP})} = 6.23 + 0.01$ (68). The error limits are as defined in Table VIII.

[b]From Table VIII.

[c]Calculated with the log $K^M_{M(\varepsilon\text{-ATP})}$ values given in (69, 70), and the values listed in Table II for log $K^M_{M(PNTP)}$ (= log $K^M_{M(\varepsilon\text{-ATP})_{op}}$); see also (29). The acidity constants of $H_2(\varepsilon\text{-ATP})^{2-}$ are $pK^H_{H_2(\varepsilon\text{-ATP})} = 4.45 \pm 0.02$ and $pK^H_{H(\varepsilon\text{-ATP})} = 6.50 \pm 0.01$ (69). The error limits are as defined in Table II.

[d]From Table II.

since binding of these two metal ions to the phosphate groups clearly dominates. However even in these cases great care should be exercised, as the results for Mg(ATP)$^{2-}$ and Mg(ε-ATP)$^{2-}$ (Table IX) indicate.

Consideration of the results of Table IX, together with preliminary results (25) for the extent of base-backbinding in M(ADP)$^-$ complexes, leads to the conclusion (35) that the total extent of macrochelate formation as calculated from stability constant comparisons apparently depends, for all metal ions studied, on the number of phosphate groups: i.e., % M(AMP)$_{cl/tot}$ < % M(ADP)$^-_{cl/tot}$ > % M(ATP)$^{2-}_{cl/tot}$. For the complexes of the ε-adenine nucleotides no final conclusions can yet be drawn in this respect, but the situation may well be similar (29).

Effect of N-7 Basicity on the Extent of Nucleic Base-Backbinding. Comparison of Macrochelate Formation in M(GMP) and M(AMP) Species

It was shown above that the stability of phosphate-metal ion complexes depends on the basicity of the coordinating phosphate group. We also have seen that base-backbinding in M(AMP) complexes occurs with N-7. Hence, one may also expect that the degree of formation of the macrochelate (Equation 3) might be influenced by the basicity of the purine N-7.

The pK$_a$ values for the H$^+$(N-7) units in three purine nucleosides are listed in Table X (65, 71). The acidity constant of H$^+$(N-7) for H(guanosine)$^+$ and H(inosine)$^+$ may be determined by direct experiment since the proton is located at N-7, while the pK$_a$ for H(N-7/adenosine)$^+$ is only indirectly accessible via estimations because the most basic site of adenosine is N-1 (65, 71). However, it is evident that the basicity of N-7 changes in the series adenosine $\lesssim$ inosine < guanosine, with the N-7 of guanosine clearly the most basic. Consequently, one would predict that due to a more intense M–N-7 interaction the GMP^{2-} complexes are more stable than the corresponding AMP^{2-} complexes (see Figure 6 for the structures); in fact, this is observed (72, 73).

Some of the equilibrium constants (64, 73) are plotted in Figure 8. It is immediately evident that Cd(GMP) is more stable than Cd(AMP), in agreement with the above reasoning. However, both complexes are significantly more stable than expected solely on the basis of the basicity of the phosphate groups, i.e. for both systems the experimental data points are clearly above the reference line. In accordance with the conclusions presented above, the value for Cd(TuMP) fits on the line. In the case of the corresponding Ca^{2+} systems only the point due to Ca(GMP) is marginally above the line, indicating a slight stability increase of this species.

A more comprehensive comparison is possible using the data collected in Table XI. The stability increase, log Δ, and the percentage of M(NMP)$_{cl}$ are larger with GMP^{2-} than with AMP^{2-} for all ten metal ions studied. This increased stability is attributed largely to the greater basicity

Table X. Acid-base Properties of N-7 in 'Neutral' Purine-Nucleosides

Guanosine:	$pK_{H(Guo)}^{H}$	$= 2.11 \pm 0.04$[a]	
Inosine:	$pK_{H(Ino)}^{H}$	$= 1.06 \pm 0.15$[a]	
Adenosine:[b]	$pK_{H(N-7/Ado)}^{H}$	~ 1.1	(71)
		$\sim 0.2 \pm 0.5$	(65)

[a] I = 0.1 M, NaNO$_3$; 25 °C. Sigel, H., 1988, results to be published.
[b] The difficulty to determine a pK_a value for H$^+$(N-7) arises from the larger proton affinity of N-1 (71).

Table XI. Comparison of the Stability Increase Due to N-7 Backbinding (log Δ; Equation 5) and the Extent of Macrochelate Formation (Equation 3) in M(GMP) and M(AMP) Complexes (25 °C; I = 0.1 M, NaNO$_3$)[a]

M^{2+}	M(GMP)		M(AMP)	
	log Δ	% M(GMP)$_{cl}$	log Δ	% M(AMP)$_{cl}$
Mg^{2+}	0.13 ± 0.04	26 ± 8	0.03 ± 0.04	7 ± 9
Ca^{2+}	0.07 ± 0.05	15 ± 10	0.00 ± 0.05	0
Sr^{2+}	0.11 ± 0.05	22 ± 10	0.00 ± 0.05	0
Ba^{2+}	0.16 ± 0.05	31 ± 9	0.01 ± 0.05	0
Mn^{2+}	0.22 ± 0.07	40 ± 10	0.07 ± 0.07	15 ± 14
Co^{2+}	0.77 ± 0.07	83 ± 3	0.29 ± 0.07	49 ± 9
Ni^{2+}	1.18 ± 0.06	93 ± 1	0.54 ± 0.06	71 ± 4
Cu^{2+}	0.72 ± 0.09	81 ± 4	0.27 ± 0.08	46 ± 10
Zn^{2+}	0.56 ± 0.08	72 ± 5	0.26 ± 0.11	45 ± 13
Cd^{2+}	0.53 ± 0.06	70 ± 4	0.24 ± 0.06	43 ± 8

[a] See Figure 6. The acidity constants of H$_2$(GMP)$^{\pm}$ are $pK_{H_2(GMP)}^{H}$ = 2.43 $\pm$ 0.05 and $pK_{H(GMP)}^{H}$ = 6.25 $\pm$ 0.02 (73). The GMP data are from (72, 73); those for AMP are from Table VIII. The error limits are as defined in Table VIII.

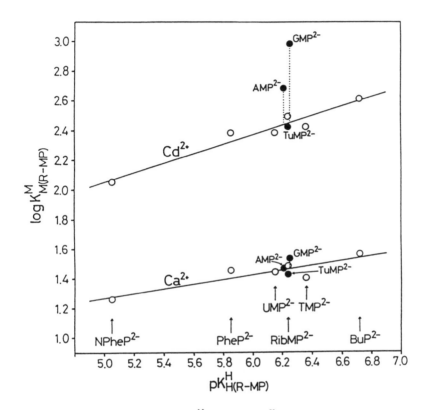

Figure 8. Relationship between log $K^M_{M(R-MP)}$ and $pK^H_{H(R-MP)}$ for the 1:1 complexes of Ca^{2+} and Cd^{2+} with some simple phosphate monoester ligands, R-MP^{2-} (O). The points due to the complexes formed with AMP^{2-}, GMP^{2-}, and $TuMP^{2-}$ (●) are inserted for comparison; the corresponding equilibrium constants are from (64, 73). For further details see the legend to Figure 7. This is an altered and, by the M(TuMP) data, expanded version of a figure published earlier (73).

of N-7 in the guanine residue compared with that in the adenine residue, although one could argue that the carbonyl oxygen at C-6 also participates in the binding process. Indeed, preliminary results with IMP^{2-} (Massoud, S. S., Sigel, H., results to be published), in which N-7 has a similar or slightly greater basicity as in AMP^{2-} (see Table X), indicate that the M(IMP) complexes are somewhat more stable than the corresponding M(AMP) species. This stability increase might possibly be partially attributed to an effect of O-6, which occurs probably *via* a hydrogen bond to this carbonyl oxygen from a metal ion-coordinated water molecule. Such structural arrangements are known from X-ray analysis of M(IMP) and M(GMP) complexes (*74, 75*). However, this effect should not be overemphasized, in agreement with conclusions for Ni^{2+}, Cu^{2+}, or Zn^{2+} binding to inosine and guanosine (*39*): "There is little if any increase in stability due to hydrogen bond formation from coordinated water to O-6 of inosine and guanosine." Hence, the overall conclusion for the present comparison is that the increased basicity of N-7 in GMP^{2-} is crucial for the enhanced complex stability of the M(GMP) species, compared with the M(AMP) complexes (Table XI).

Structural Considerations for Macrochelates Formed with Purine Nucleoside 5'-Monophosphates

In focusing on N-7 (without denying an indirect role of O-6) one may propose the following four different structures for the macrochelates formed with purine nucleoside 5'-monophosphates in M(NMP) species: (i) The metal ion coordinates in an inner sphere manner to N-7, and outer sphere, i.e. *via* water, to the phosphate group. Such structures are known from X-ray analysis (*74–76*), and N-7 inner sphere coordination has also been demonstrated by ultraviolet spectral perturbation studies (*43, 77, 78*). (ii) The metal ion coordinates only outer sphere to N-7, but inner sphere to the phosphate group (this is in analogy to Figure 3B). (iii) Both binding sites, N-7 and the phosphate group, coordinate inner sphere to the metal ion. (iv) A structure in which both binding sites interact exclusively in an outer sphere mode with the metal ion is expected to play only a minor role (i.e., at best for the alkaline earth ions with, e.g., GMP^{2-}). The binding modes for the open M(NMP) species involving only phosphate binding (Equation 3) are those discussed above.

Structures (i) and (ii) agree with the suggestion (*38, 43, 79*) that steric hindrance "reduces the opportunity for both an α-phosphate and N-7 to be coordinated inner sphere in a macrochelate at the same time. One of the coordinate bonds is outer sphere: either N-7 inner- and α-phosphate outer sphere or N-7 outer sphere and α-phosphate inner sphere" (*79*), but the entirely inner sphere structure (iii) does not. However, there are indications from kinetic studies that both phosphate and base may bind inner sphere to the metal ion (*80–84*), and space-filling molecular models allow the construction of such a macrochelate without strain (*68*). In addition, kinetic and product studies (*85, 86*) for the reaction between *cis*-Pt(NH$_3$)$_2^{2+}$ and 5'-(2'-deoxy)GMP^{2-} indicate a direct coordination of the metal ion to N-7 and the phosphate group. This structure was independently confirmed in a careful NMR study (*87*) for *cis*-Pt(NH$_3$)$_2^{2+}$ and IMP^{2-} and GMP^{2-} (see also *44*). Furthermore, there is now also evidence (*88*) that in D$_2$O the Mo(IV) of the (η^5-C$_5$H$_5$)$_2$Mo^{2+} unit coordinates directly to N-7 and the phosphate group of 5'-(2'-deoxy)AMP^{2-} forming a macrochelate. Moreover, it is difficult to see how such substantial stability increases (log Δ; Equation 5) as seen in Tables IX and XI could arise *via* outer sphere interactions only. Hence, the formation of entirely inner sphere complexes among the M(NMP) species appears certain. One should add, though, that the results of model-building differ with nucleoside 5'-di- and -tri-phosphates: inner sphere coordination of N-7 and of the β and γ phosphate groups makes direct coordination of the α group more difficult. Here the steric situation is facilitated by a water molecule between the metal ion and the α phosphate group in accord with previous suggestions (*38, 43, 79*).

To conclude, the percentages calculated and presented above (Tables VIII, IX and XI) for the degree of formation of the macrochelated form (Equation 3) in the M(NMP) systems encompass structures (i), (ii), and (iii) [and in the case of the alkaline earth ions and GMP^{2-}, possibly also (iv)]; i.e., the listed values have to be considered as % M(NMP)$_{cl/tot}$ (*64*). Obviously, the ratios of these macrochelated isomers will differ from nucleotide to nucleotide and from metal ion to metal ion. For example, for Cu(AMP) (*64*), Cu(GMP), and Ni(GMP) substantial amounts of the macrochelate with structure (iii) are expected, Ni(AMP)$_{cl}$ is probably a mixture of structures (i) and (iii) (*64*), while for the M(GMP) complexes of Mg^{2+}, Ca^{2+}, Sr^{2+}, and Ba^{2+} structure (ii) appears to be the most probable one.

Influence of the Position of the Phosphate Group on the Ribose Ring on the Nucleic Base-Metal Ion Interaction. Comparison of the Complexing Properties of 2'-AMP^{2-}, 3'-AMP^{2-} and 5'-AMP^{2-}

Adenosine 5'-monophosphate in its preferred *anti* conformation (Figure 6) exists in an ideal orientation for the simultaneous coordination of a metal ion to the phosphate group and to N-7 (see above). To learn if a metal ion-base interaction is also possible with the phosphate group in a less favored position, the coordinating properties of 2'-AMP^{2-} and 3'-AMP^{2-} were studied (*89, 90*) and compared with those of 5'-AMP^{2-} (Figure 9).

Figure 10 shows the results obtained for the Zn^{2+} and Cu^{2+} systems with the three AMPs. It is evident that with Cu^{2+} an increased stability is observed not only for the complex with 5'-AMP^{2-}, but clearly also for that with 2'-AMP^{2-} and possibly marginally with 3'-AMP^{2-}. In any case, this

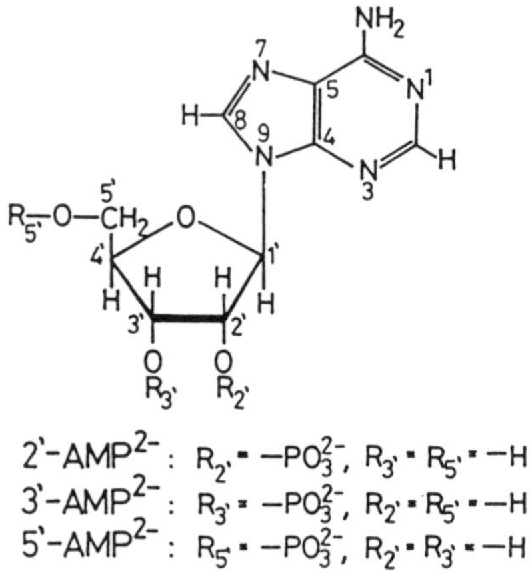

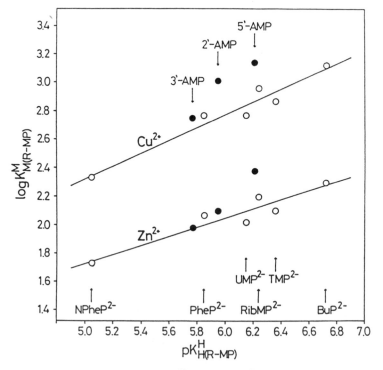

$2'$-AMP^{2-} : $R_{2'}$ = $-PO_3^{2-}$, $R_{3'}$ = $R_{5'}$ = $-H$
$3'$-AMP^{2-} : $R_{3'}$ = $-PO_3^{2-}$, $R_{2'}$ = $R_{5'}$ = $-H$
$5'$-AMP^{2-} : $R_{5'}$ = $-PO_3^{2-}$, $R_{2'}$ = $R_{3'}$ = $-H$

Figure 9. Chemical structures of adenosine monophosphates (AMPs) in their dominating anti conformations (27).

Figure 10. Relationship between log $K^M_{M(R-MP)}$ and p$K^H_{H(R-MP)}$ for the 1:1 complexes of Cu^{2+} and Zn^{2+} with some simple phosphate monoester ligands, R-MP^{2-} (O). The points due to the complexes formed with $2'$-AMP^{2-}, $3'$-AMP^{2-} and $5'$-AMP^{2-} (●) are inserted for comparison; the corresponding equilibrium constants are from (64, 89). See also the legend to Figure 7.

indicates that intramolecular interactions of a phosphate-coordinated metal ion are possible not only with N-7 but also with other sites. Among the Zn^{2+} systems, the situation is unequivocal only for the 5'-AMP^{2-} complex and marginal for the 2'-AMP^{2-} species; the stability of Zn(3'-AMP) is clearly determined by the coordinating properties of the phosphate group.

The analysis of the equilibrium constants for 10 metal ion systems with the three AMPs is summarized in Table XII (89). The indicated observations for Cu(2'-AMP) are confirmed: in this case significant amounts of macrochelate exist. For the 2'-AMP^{2-} complexes with Ni^{2+}, Zn^{2+}, and Cd^{2+} the observed stability increase (log Δ) is just at the limit of significance. Taking into account that the properties of 2'-AMP^{2-} and 3'-AMP^{2-} are not analogous with regard to their complexing abilities, it is evident that the increased stability observed for the M(2'-AMP) complexes cannot just be due to an interaction with the neighboring 3'-OH group, because exactly the same type of interaction would be possible in M(3'-AMP) complexes with the 2'-OH group. Consideration of the structure shown in Figure 9 indicates that the only other site that may be reached by a metal ion coordinated to the phosphate group in the 2' position is N-3 of the adenine residue. Indeed, such an interaction has previously been suggested for the Cu(2'-AMP) system based on ^{1}H NMR line broadening studies (91, 92). Hence, one must conclude that in M(2'-AMP) systems, macrochelates form involving the phosphate group and N-3 [although a smaller fraction of 7-membered chelates involving the 3'-OH might also exist (see below)].

Among the M(3'-AMP) complexes (Table XII) those containing Ni^{2+} and Cu^{2+} are just at the limit of a significant stability increase (log Δ; Equation 5). However, the occurrence of a further interaction besides the phosphate coordination is certain. With Cu^{2+} experiments were also carried out in 50% (v/v) dioxane-water as solvent. In this case the stability increase for Cu(3'-AMP) is rather pronounced (Liang, G.; Sigel, H., results to be published). There are two possible explanations for this observation: (i) The 3' phosphate-coordinated metal ion interacts with the neighboring 2'-OH group. The consequence of this assumption is that a small fraction of the stability increase observed for M(2'-AMP) species should then also be attributed, due to the symmetric situation, to an interaction between the 2' phosphate-coordinated metal ion and the 3'-OH group. (ii) In a fraction of the M(3'-AMP) species the ligand adopts the less favored *syn* conformation. A metal ion coordinated to the 3' phosphate group may then also interact with N-3. The latter explanation (ii) is thought to be more important (though species corresponding to those in point (i) could still be formed in small amounts), because the energy barrier between the *anti* and *syn* conformations is smaller with 3'-AMP (and 2'-AMP) than with 5'-AMP (93), and there are indications from an early ultraviolet difference-spectral study (77) that a Cu–base interaction occurs in Cu(3'-AMP) species.

Table XII. Comparison of the Stability Increase (log Δ; Equation 5)
Observed in Different M(AMP) Complexes and Extent of the
Corresponding Macrochelate Formation (Equation 3) (25 °C; I = 0.1 M,
NaNO$_3$)[a]

	M(2'-AMP)		M(3'-AMP)		M(5'-AMP)	
M^{2+}	log Δ	% M(AMP)$_{cl}$	log Δ	% M(AMP)$_{cl}$	log Δ	% M(AMP)$_{cl}$
Mg^{2+}	0.02±0.06	0	0.02±0.05	0	0.03±0.04	7±9(≤15)
Ca^{2+}	0.01±0.05	0	–0.03±0.05	0	0.00±0.05	0
Sr^{2+}	–0.02±0.05	0	–0.05±0.06	0	0.00±0.05	0
Ba^{2+}	–0.02±0.06	0	–0.05±0.06	0	0.01±0.05	0
Mn^{2+}	0.05±0.07	11+15(≤24)	0.01±0.07	0	0.07±0.07	15±14
Co^{2+}	0.05±0.08	11±16(≤24)	–0.04±0.08	0	0.29±0.07	49±9
Ni^{2+}	0.06±0.07	13±13(≤25)	0.06±0.07	13±13(≤25)	0.54±0.06	71±4
Cu^{2+}	0.26±0.08	45±10	0.08±0.08	17±16(≤30)	0.27±0.08	46±10
Zn^{2+}	0.07±0.08	15±17(≤29)	0.00±0.08	0	0.26±0.11	45±13
Cd^{2+}	0.06±0.07	13±13(≤25)	0.02±0.06	0	0.24±0.06	43±8

[a]See Figure 9. The acidity constants of $H_2(2'\text{-AMP})^{\pm}$ are $pK^H_{H_2(AMP)} = 3.74 \pm 0.02$ and $pK^H_{H(AMP)} = 5.95 \pm 0.01$, and those of $H_2(3'\text{-AMP})^{\pm}$ are $pK^H_{H_2(AMP)} = 3.70 \pm 0.04$ and $pK^H_{H(AMP)} = 5.77 \pm 0.02$ (89). The data for the 2'-AMP^{2-} and 3'-AMP^{2-} systems are from (89), those for 5'-AMP are from Table VIII. The error limits are as defined in Table VIII.

The observations in 50% dioxane-water mentioned above should be kept in mind, as it is now well established that the 'equivalent solution' or 'effective' dielectric constant in the active site cavity of an enzyme is reduced compared to the dielectric constant of bulk water (94–97). Hence, under such conditions certain groups, which would be of insignificant importance in aqueous solution, might become metal ion-binding sites .

The acid-base properties (27) and especially the self-stacking tendencies (26) of the three AMPs are rather similar, but these nucleotides develop individual features in the presence of metal ions. Considering the results assembled in Table XII as a whole it is evident that in aqueous solution among the three AMPs the 5' isomer forms the largest quantities of macrochelates in M(AMP) systems. Since N-3 seems to be about as easily accessible to a metal ion bound to the 2' phosphate group as is N-7 to a metal ion bound to the 5' phosphate group, one may suggest that N-7 of the adenine residue has a greater affinity for metal ions than N-3. Taking into account recent comparisons (65) of the coordinating properties of the imidazole-like N-7 and the pyridine-like N-1, which is sterically somewhat hindered by the 6-NH$_2$ group of the adenine residue, the order of decreasing affinity for many metal ions tentatively appears to be N-7 $\gtrsim$ N-1 $\gtrsim$ N-3. That N-3 of the purine moiety is able to coordinate metal ions is also known from X-ray crystal structure studies of Pt(II) complexes of guanine derivatives (98), as well as of Rh(I) complexes of 8-azaadenine derivatives (99).

Intramolecular Stacking or Hydrophobic Ligand-Ligand Interactions in Mixed-Ligand Complexes Containing Nucleotides

The formation of mixed-ligand complexes provides a further dimension in the aim to achieve, to observe, and to quantify selective interactions. The two ligands in a ternary M(A)(B) complex may interact indirectly with each other, i.e. only *via* the commonly coordinated metal ion (15, 100, 101), or they may undergo direct interactions forming covalent bonds, and intramolecular ion pairs, stacks, or hydrophobic adducts (15, 33, 34, 102, 103). Clearly, such direct interactions require the presence of suitable groups in the coordinated ligands.

In the remainder of this section a few examples of intramolecular hydrophobic and aromatic ring stacking interactions, as shown in the equilibria of Equations 10 and 11, will be considered:

$$\tag{10}$$

indole — M ... O—P (phosphate) ... N ... O—P ... ribose ... purine ⇌ M ... O—P (phosphate) ... N ... indole ... purine—ribose — O—P   (11)

If the 'open' form is designated as $M(L)(NP)_{op}$, and the species with the intramolecular adduct as $M(L)(NP)_{ad}$, the dimensionless equilibrium constant $K_{I/ad}$ is defined by Equation 12:

$$K_{I/ad} = \frac{[M(L)(NP)_{ad}]}{[M(L)(NP)_{op}]} \qquad (12)$$

In the present case NP represents a nucleotide, and L either 2,2'-bipyridyl (Bpy), 1,10-phenanthroline (Phen), $(D),L$-tryptophanate (Trp$^-$), or L-leucinate. The equilibrium constants $K_{I/ad}$ and the percentages for $M(L)(NP)_{ad}$ of some mixed-ligand–metal ion systems are collected in Table XIII. Such data may help to improve our understanding of intercalation reactions, e.g., between DNA and complexes formed with Phen, Bpy or related ligands (104–107). In all the systems considered in Table XIII a possibly previously existing nucleic base–metal ion interaction in the binary M(NP) complexes is released upon the formation of the mixed-ligand complexes (see above).

The results presented in Table XIII allow many comparisons. A few are indicated: (i) The size of the aromatic base residue of the nucleotide has, as one would expect, an influence on the degree of formation of the intramolecular adduct: purines stack better than pyrimidines (cf. No. 1 & 2, 4 & 5, or 11–14). (ii) The steric opportunities of the nucleotide influence the extent of the interaction: the degree of formation of $Cu(Phen)(AMP)_{ad}$ differs with 2'-AMP^{2-}, 3'-AMP^{2-}, and 5'-AMP^{2-} (see Figure 9, and No. 15–17). (iii) The degree of formation of an intramolecular adduct, e.g. of $M(Phen)(ATP)_{ad}^{2-}$, may be quite comparable for several metal ions (provided the ligands are flexible enough to overcome the restrictions of the metal ion coordination spheres; No. 8–10), despite the fact that the overall stabilities of the ternary $M(Phen)(ATP)^{2-}$ complexes with Mg^{2+}, Ca^{2+}, and Cu^{2+} are very different (108). (iv) Intramolecular aromatic ring stacks are more stable than simple hydrophobic adducts (33, 102); e.g., a purine-indole interaction is more pronounced than a purine-isopropyl one, as is indicated from the $M(ATP)(Trp)^{3-}$ and $M(ATP)(Leu)^{3-}$ examples (No. 21–26). (v) Comparison of the $K_{I/ad}$ values and the degrees of

Table XIII. Dimensionless Equilibrium Constants $K_{I/ad}$ (Equation 12)[a] and Degree of Formation of the Intramolecular Hydrophobic and Stacking Adducts in Some Ternary M(L)(NP) Complexes (Equations 10, 11)[a], i.e. %M(L)(NP)$_{ad}$

No.	M(L)(NP)	$K_{I/ad}$ Pot.[b]	$K_{I/ad}$ NMR[c]	%M(L)(NP)$_{ad}$ Pot.[b]	%M(L)(NP)$_{ad}$ NMR[c]	Ref
1	Cu(Bpy)(5'-UMP)	~0.5		~30		68
2	Cu(Bpy)(5'-AMP)	~4.0		~80		68
3	Cu(Bpy)(ε-AMP)	~4.0		~80		68
4	Cu(Bpy)(5'-UTP)$^{2-}$	~1.2		~55		34
5	Cu(Bpy)(5'-ATP)$^{2-}$	5.9		86 ± 3		34
6	Cu(Bpy)(ε-ATP)$^{2-}$	~10		~90		69
7	Cu(Phen)(5'-ATP·H)$^-$	~3.8		~79		34
8	Cu(Phen)(5'-ATP)$^{2-}$	11		92 ± 2		34
9	Mg(Phen)(5'-ATP)$^{2-}$	~10		~90	~100	15,55
10	Ca(Phen)(5'-ATP)$^{2-}$	~10		~90		15
11	Zn(Bpy)(5'-UTP)$^{2-}$	~1.8	~0.7	~65	~40	42,55
12	Zn(Bpy)(5'-ATP)$^{2-}$	~2.3	~1.2	~70	~55	42,55
13	Cd(Bpy)(5'-CTP)$^{2-}$	~1.2		~55		42
14	Cd(Bpy)(5'-ATP)$^{2-}$	~1.5		~60		42
15	Cu(Phen)(2'-AMP)	35.3 ± 6.7		97.2 ± 0.5		90[d]
16	Cu(Phen)(3'-AMP)	1.29 ± 0.45		56 ± 9		90[d]
17	Cu(Phen)(5'-AMP)	8.77 ± 1.81		90 ± 2		d
18	Mg(Trp)(ε-ATP)$^{3-}$		1.0		51 + 11	69
19	Zn(Trp)(ε-ATP)$^{3-}$		0.45		31 ± 6	69
20	Mg(Trp)(ATP)$^{3-}$		0.8		44 ± 19	69
21	Mn(Trp)(ATP)$^{3-}$	1.09 ± 0.52		52 ± 12		33[e]
22	Cu(Trp)(ATP)$^{3-}$	0.55 ± 0.33		35 ± 14		33[e]
23	Zn(Trp)(ATP)$^{3-}$	2.89 ± 0.55	0.7	74 ± 4	40 ± 15	33[e],69
24	Mn(Leu)(ATP)$^{3-}$	0.7 ± 0.9		41 ± 31		33[e]
25	Cu(Leu)(ATP)$^{3-}$	0.26 ± 0.22		21 ± 14		33[e]
26	Zn(Leu)(ATP)$^{3-}$	0.05 ± 0.17		5 ± 15	~30(20/75)[f]	33[e]

[a] In many cases assumptions had to be made in the calculations and therefore only estimates are obtained. ^{1}H NMR shift experiments provide direct evidence that an adduct is formed, but %M(L)(NP) is usually only an estimate. The error limits given are 3 times the standard error of the mean or the sum of the probable systematic errors, whichever is larger. For details the original references should be consulted.

[b] Based on potentiometric pH titrations of aqueous solutions; I = 0.1 M, NaNO$_3$ or NaClO$_4$; 25 °C.

[c] Based on ^{1}H NMR shift experiments in D$_2$O; I = 0.1 M, NaNO$_3$; 27 °C.

[d] Massoud, S. S.; Tribolet, R.; Sigel, H., details to be published.

[e] The error limits have now been calculated with the data given in (33).

[f] The lower and upper limits are given in parenthesis.

formation of $M(L)(NP)_{ad}$ species in Table XIII with the information given in Table IV reveals that the energy changes (ΔG^0) associated with Equations 10 and 11 are usually small.

It should be emphasized that the formation of a metal ion bridge between two ligands with suitable groups for an interaction makes this interaction strongly favorable (33, 34, 55, 103, 109). For example, corresponding unbridged stacks or hydrophobic adducts are less stable and have a lower degree of formation under given conditions in the absence of a bridging metal ion (29, 34, 110). Moreover, unbridged binary adducts are very sensitive to the addition of organic solvents like ethanol or dioxane to an aqueous solution containing such adducts (34, 111). The behavior of ternary metal ion-bridged intramolecular adducts is quite different: they are much less sensitive (34), and their degree of formation may even be favored (111, 112).

Caveat and General Conclusions

Intramolecular equilibria in binary and ternary mixed-ligand complexes of nucleotides, as exemplified by Equations 3, 10 and 11, often involve only small changes in free energy (ΔG^0). Formation of 20% of a certain species in rapid equilibrium may be more than enough for a given enzymic reaction even though the energy change, $\Delta G^0 = -0.6$ kJ/mol, is very small (Table IV). It is evident that Nature has here a tool to achieve high selectivity *via* the involvement of preferred structures simply by connecting several such equilibria in a cascade-like way without falling into a deep energy valley, which is not desirable for enzymic turnover processes.

The examples of metal ion-nucleotide systems discussed in this chapter provide information on the structures of M(NP) complexes one must consider in enzymic reactions. For example, $M(ATP)^{2-}$ complexes may exist at least in a 'simple' phosphate coordinated form and in two macrochelated forms, one interacting inner sphere and the other outer sphere, *via* a coordinated water molecule, with N-7 of the adenine residue (Figure 3). It is evident that each of these three main species can occur in further isomeric forms, differing, e.g., in their phosphate-coordinating mode or their arrangement in an octahedral coordination sphere. That seemingly small alterations may have significant effects on the reactivity of a system was shown in studies of the metal ion-promoted transfer of the terminal phosphate group of nucleoside 5'-triphosphates to water (59, 113).

The results summarized here also show that certain sites in nucleic base residues have a greater affinity for metal ions than others. For example, one may predict where particular 3d transition metal ions, as well as Zn^{2+} and Cd^{2+}, bind to DNA. Due to the Watson-Crick base pairing of nucleic base residues only a limited number of possible binding sites is available, as is evident from Figure 11. The most basic and

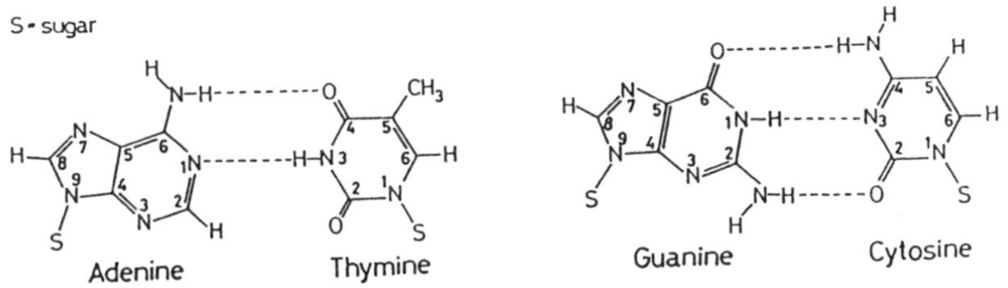

Figure 11. Chemical structures of the Watson-Crick base-pairs in DNA.

therefore preferred metal ion-binding site is clearly N-7 of the guanine residue, followed by N-7 of the adenine moiety, and only then by the N-3 atoms. Binding studies of cis-$[Pt(NH_3)_2]^{2+}$ to DNA confirm this order: N-7 of the guanine residue is the preferred binding site (114–117).

Opening of the double helix leading to single-stranded DNA gives metal ions access to two further binding sites, N-1 of the adenine moiety and N-3 of the cytosine residue. The same sites are available in single-stranded RNA. Both sites are sterically somewhat shielded due to the *ortho* amino group and therefore their metal ion affinity is lower than would be expected on the basis of their basicity (65). For divalent 3d transition metal ions, Zn^{2+}, and Cd^{2+}, the cytosine N-3 site is preferred to the adenine N-1 site. N-1 and N-7 of the adenine residue have a similar affinity, though that of N-7 is slightly greater towards Ni^{2+}, Cu^{2+}, and Zn^{2+} (65). One may therefore tentatively propose, in the neutral pH range, the following overall order of affinity of the metal ion-binding sites available in single-stranded nucleic acids: N-7/Guo $\gtrsim$ N-3/Cyt $\gtrsim$ N-7/Ado $\gtrsim$ N-1/Ado > N-3/Ado,Guo. Some of the metal ions, e.g. Cu^{2+}, may also be able in the neutral pH range to partially replace H^+ from the neutral H(N-1) unit in the guanine moiety (cf. Figure 11) or from H(N-3) of the uracil or thymine residues (24, 39, 63). The affinities of these N^- sites must then be inserted into the preceding series at various places.

It is interesting to consider in this connection some previous studies on the influence of H_2O_2 on DNA in the presence of Cu^{2+}. Native DNA coordinates Cu^{2+} in neutral or slightly alkaline aqueous solution, preferentially *via* the phosphate groups. Addition of H_2O_2 gives rise to a catalase-like activity, as well as to the formation of ternary peroxo complexes as observed by spectrophotometry (118). There is no evident degradation of the nucleic bases under these conditions. However, if Cu^{2+} is added to DNA and the solution is kept at room temperature for 1.5 days, Cu^{2+} penetrates into DNA and coordinates to the base residues. Addition of H_2O_2 to such a solution leads not only to a catalase-like activity but also to a peroxidase-like one. Now the nucleic bases are affected, as can be followed by measuring the absorption at 260 nm (119). It is well known that Cu^{2+} may bind to the bases of DNA (120, 121), and by the indicated experiments (119) native and denatured DNA can be distinguished. RNA, with its less complete base-pairing, offers even in the native form nucleic base sites for Cu^{2+} binding, and consequently base degradation occurs in the presence of H_2O_2 (119).

These observations are the result of the fact that the reactions do not proceed via free HO• radicals, but occur in the coordination sphere of Cu^{2+} (122–124). Consequently, the Cu^{2+}/H_2O_2 system can be used to probe the structures of complexes in solution (125), including those of macromolecules. For example, in this way it may be shown that the double helix of DNA is rather stable in the pH range 6 to 10, but is opened at pH < 5 and pH > 11. The stabilizing effect of the alkali metal ions and Mg^{2+} on the double helix may in this way also be characterized (126).

Similarly, evidence for base-backbinding in monomeric Cu^{2+} complexes of nucleoside 5'-triphosphates was provided by this method many years ago (*127*). Quantification of the peroxidase-like activity in such systems leads to bell-shaped curves. In the neutral pH range the activity increases with increasing pH, and then decreases at higher pH, due to the formation of hydroxo complexes. As discussed above, e.g., coordination of OH^- to $M(ATP)^{2-}$ species releases adenine N-7 from the coordination sphere, and consequently in the $Cu^{2+}/ATP/H_2O_2$ system the oxidation of the base residue by the Cu^{2+}-peroxo unit is inhibited. Part of this Cu^{2+}/H_2O_2 work has been summarized (*128*).

Another point of interest in the present context is the interaction between nucleic base moieties and amino acid side chains (*129, 130*), as such interactions usually occur between nucleic acids and proteins (*131*) in the presence of metal ions (*132*). From the results summarized above it follows that the affinity of $M(ATP)^{2-}$ for the following amino acids decreases in the order tryptophanate > leucinate > alaninate. Such results are meaningful with regard to the origin of the genetic code (*133*), as well as for the prediction of preferred intercalation reactions, e.g., with DNA. That aromatic amino acid side chains, e.g., the indole moiety of a Trp residue, may function in nucleic acid binding *via* intercalation is well known (*134*), and that aliphatic amino acid residues may also do so, though less effectively, has repeatedly been suggested (*33, 135, 136*).

Acknowledgment

The financial support of the research of my group on nucleotide complexes by the Swiss National Science Foundation is gratefully acknowledged.

Literature Cited

1. Boyer, P. D. *Biochemistry* **1987**, *26*, 8503–8507.
2. Westheimer, F. H. *Science* **1987**, *235*, 1173–1178.
3. Cohn, M. *Acc. Chem. Res.* **1982**, *15*, 326–332.
4. Szent-Györgyi, A. In "Enzymes: Units of Biological Structure and Function"; Gaebler, O. H., Ed.; Academic Press: New York, 1956; p 393.
5. Cohn, M.; Hughes, T. R., Jr. *J. Biol. Chem.* **1962**, *237*, 176–181.
6. Cohn, M.; Hughes, T. R., Jr. *J. Biol. Chem.* **1960**, *235*, 3250–3253.
7. Schneider, P. W.; Brintzinger, H., Erlenmeyer, H. *Helv. Chim. Acta* **1964**, *47*, 992–1002.
8. Brintzinger, H. *Helv. Chim. Acta* **1961**, *44*, 935–939.
9. Brintzinger, H. *Biochim. Biophys. Acta* **1963**, *77*, 343–345.

10. Schneider, P. W.; Brintzinger, H. *Helv. Chim. Acta* **1964**, *47*, 1717–1733.
11. Sternlicht, H.; Shulman, R. G.; Anderson, E. W. *J. Chem. Phys.* **1965**, *43*, 3123–3132.
12. Sternlicht, H.; Shulman, R. G.; Anderson, E. W. *J. Chem. Phys.* **1965**, *43*, 3133–3143.
13. Sigel, H.; Becker, K.; McCormick, D. B. *Biochim. Biophys. Acta* **1967**, *148*, 655–664.
14. Naumann, C. F.; Sigel, H. *J. Am. Chem. Soc.* **1974**, *96*, 2750–2756.
15. Sigel, H. In "Coordination Chemistry–20"; Banerjea, D., Ed.; published by IUPAC via Pergamon Press: Oxford and New York, 1980; pp 27–45.
16. Phillips, R., S. *J. Chem. Rev.* **1966**, *66*, 501–527.
17. Ts'o, P. O. P.; Melvin, I. S.; Olson, A. C. *J. Am. Chem. Soc.* **1963**, *85*, 1289–1296.
18. Chan, S. I.; Schweizer, M. P.; Ts'o, P. O. P.; Helmkamp, G. K. *J. Am. Chem. Soc.* **1964**, *86*, 4182–4188.
19. Mitchell, P. R.; Sigel, H. *Eur. J. Biochem.* **1978**, *88*, 149–154.
20. Neurohr, K. J.; Mantsch, H. H. *Can. J. Chem.* **1979**, *57*, 1986–1994.
21. Dimicoli, J.-L.; Hélène, C. *J. Am. Chem. Soc.* **1973**, *95*, 1036–1044.
22. Heyn, M. P.; Bretz, R. *Biophys. Chem.* **1975**, *3*, 35–45.
23. Mitchell, P. R. *J. Chem. Soc. Dalton Trans.* **1980**, 1079–1086.
24. Scheller, K. H.; Hofstetter, F.; Mitchell, P. R.; Prijs, B.; Sigel, H. *J. Am. Chem. Soc.* **1981**, *103*, 247–260.
25. Scheller, K. H.; Sigel, H. *J. Am. Chem. Soc.* **1983**, *105*, 5891–5900.
26. Tribolet, R.; Sigel, H. *Biophys. Chem.* **1987**, *27*, 119–130.
27. Tribolet, R.; Sigel, H. *Eur. J. Biochem.* **1987**, *163*, 353–363.
28. Tribolet, R.; Sigel, H. *Eur. J. Biochem.* **1988**, *170*, 617–626.
29. Sigel, H. *Chimia* **1987**, *41*, 11–26.
30. Phillips, J. H.; Allison, Y. P.; Morris, S. J. *Neuroscience* **1977**, *2*, 147–152.
31. Winkler, H.; Westhead, E. *Neuroscience* **1980**, *5*, 1803–1823.
32. Winkler, H.; Carmichael, S. W. In "The Secretory Granule"; Poisner, A. M.; Trifaró, J. M.; Eds.; Elsevier Biomedical Press: Amsterdam, New York and Oxford, 1982; pp 3–79.
33. Sigel, H.; Fischer, B. E.; Farkas, E. *Inorg. Chem.* **1983**, *22*, 925–934.
34. Tribolet, R.; Malini-Balakrishnan, R.; Sigel, H. *J. Chem. Soc. Dalton Trans.* **1985**, 2291–2303.
35. Sigel, H. *Eur. J. Biochem.* **1987**, *165*, 65–72.
36. Sigel, H.; Tribolet, R.; Malini-Balakrishnan, R.; Martin, R. B. *Inorg. Chem.* **1987**, *26*, 2149–2157.
37. Martin, R. B.; Sigel, H. *Comments Inorg. Chem.* **1988**, *6*, 285–314.
38. Martin, R. B.; Mariam, Y. H. *Met. Ions Biol. Syst.* **1979**, *8*, 57–124.

39. Martin, R. B. *Acc. Chem. Res.* **1985**, *18*, 32–38.
40. Frey, C. M.; Stuehr, J. E. *J. Am. Chem. Soc.* **1972**, *94*, 8898–8904.
41. Tribolet, R.; Martin, R. B.; Sigel, H. *Inorg. Chem.* **1987**, *26*, 638–643.
42. Sigel, H.; Scheller, K. H.; Milburn, R. M. *Inorg. Chem.* **1984**, *23*, 1933–1938.
43. Mariam, Y. H.; Martin, R. B. *Inorg. Chim. Acta* **1979**, *35*, 23–28.
44. Reily, M. D.; Hambley, T. W.; Marzilli, L. G. *J. Am. Chem. Soc.* **1988**, *110*, 2999–3007.
45. Takeuchi, H.; Murata, H.; Harada, I. *J. Am. Chem. Soc.* **1988**, *110*, 392–397.
46. Onori, G. *Biophys. Chem.* **1987**, *28*, 183–190.
47. Irving, H.; Williams, R. J. P. *Nature* **1948**, *162*, 746–747.
48. Irving, H.; Williams, R. J. P. *J. Chem. Soc.* **1953**, 3192–3210.
49. Sigel, H.; McCormick, D. B. *Acc. Chem. Res.* **1970**, *3*, 201–208.
50. Sundberg, R. J.; Martin, R. B. *Chem. Rev.* **1974**, *74*, 471–517.
51. Martin, R. B. *Met. Ions Biol. Syst.* **1986**, *20*, 21–65.
52. Saha, N.; Sigel, H. *J. Am. Chem. Soc.* **1982**, *104*, 4100–4105.
53. Cini, R.; Marzilli, L. G. *Inorg. Chem.* **1988**, *27*, 1855–1856.
54. Buisson, D. H.; Sigel, H. *Biochim. Biophys. Acta* **1974**, *343*, 45–63.
55. Mitchell, P. R.; Prijs, B.; Sigel, H. *Helv. Chim. Acta* **1979**, *62*, 1723–1735.
56. Orioli, P.; Cini, R.; Donati, D.; Mangani, S. *J. Am. Chem. Soc.* **1981**, *103*, 4446–4452.
57. Sheldrick, W. S. *Angew. Chem.* **1981**, *93*, 473–474; *Angew. Chem. Int. Ed. Engl.* **1981**, *20*, 460.
58. Sheldrick, W. S. *Z. Naturforsch., B: Anorg. Chem., Org. Chem.* **1982**, *37B*, 863–871.
59. Sigel, H.; Hofstetter, F.; Martin, R. B.; Milburn, R. M.; Scheller-Krattiger, V.; Scheller, K. H. *J. Am. Chem. Soc.* **1984**, *106*, 7935–7946.
60. Massoud, S. S.; Sigel, H. *Inorg. Chem.* **1988**, *27*, 1447–1453.
61. Brintzinger, H. *Helv. Chim. Acta* **1965**, *48*, 47–54.
62. Brintzinger, H.; Hammes, G. G. *Inorg. Chem.* **1966**, *5*, 1286–1287.
63. Sigel, H. *J. Am. Chem. Soc.* **1975**, *97*, 3209–3214.
64. Sigel, H.; Massoud, S. S.; Tribolet, R. *J. Am. Chem. Soc.* **1988**, *110*, 6857–6865
65. Kinjo, Y.; Tribolet, R.; Corfù, N. A.; Sigel, H. *Inorg. Chem.* **1989**, *28*, 1480–1489.
66. Theophanides, T.; Tajmir-Riahi, H. A. *NATO ASI Ser., Ser. C* **1984**, No. *139*, 137–152.
67. Sternglanz, H.; Subramanian, E.; Lacey, J. C., Jr.; Bugg, C. E. *Biochemistry* **1976**, *15*, 4797–4802.
68. Sigel, H.; Scheller, K. H. *Eur. J. Biochem.* **1984**, *138*, 291–299.

69. Sigel, H.; Scheller, K. H.; Scheller-Krattiger, V.; Prijs, B. *J. Am. Chem. Soc.* **1986**, *108*, 4171–4178.

70. Kaden, T. A.; Scheller, K. H.; Sigel, H. *Inorg. Chem.* **1986**, *25*, 1313–1315.

71. Kim, S.-H.; Martin, R. B. *Inorg. Chim. Acta* **1984**, *91*, 19–24.

72. Massoud, S. S.; Sigel, H. in "26th International Conference on Coordination Chemistry", Porto, Portugal, **1988**; Abstract No. C49.

73. Massoud, S. S.; Sigel, H. *Bull. Chem. Soc. Ethiopia* **1988**, *2*, 9–14.

74. Gellert, R. W.; Bau, R. *Met. Ions Biol. Syst.* **1979**, *8*, 1–55.

75. Swaminathan, V.; Sundaralingam, M. *Crit. Rev. Biochem.* **1979**, *6*, 245–336.

76. Collins, A. D.; De Meester, P.; Goodgame, D. M. L.; Skapski, A. C. *Biochim. Biophys. Acta* **1975**, *402*, 1–6.

77. Sigel, H. *Experientia* **1966**, *22*, 497–499.

78. Sigel, H.; Erlenmeyer, H. *Helv. Chim. Acta* **1966**, *49*, 1266–1274.

79. Martin, R. B. *Met. Ions Biol. Syst.* **1988**, *23*, 315–330.

80. Taylor, R. S.; Diebler, H. *Bioinorg. Chem.* **1976**, *6*, 247–264.

81. Peguy, A.; Diebler, H. *J. Phys. Chem.* **1977**, *81*, 1355–1358.

82. Diebler, H. *J. Mol. Catal.* **1984**, *23*, 209–217.

83. Thomas, J. C.; Frey, C. M.; Stuehr, J. E. *Inorg. Chem.* **1980**, *19*, 505–510.

84. Frey, C. M.; Stuehr, J. E. *Met. Ions Biol. Syst.* **1974**, *1*, 51–116.

85. Evans, D. J.; Green, M.; van Eldik, R. *Inorg. Chim. Acta* **1987**, *128*, 27–29.

86. Green, M.; Miller, J. M. *J. Chem. Soc., Chem. Commun.* **1987**, 1864–1865; Corr. **1988**, 404.

87. Reily, M. D.; Marzilli, L. G. *J. Am. Chem. Soc.* **1986**, *108*, 8299–8300.

88. Kuo, L. Y.; Kanatzidis, M. G.; Marks, T. J. *J. Am. Chem. Soc.* **1987**, *109*, 7207–7209.

89. Massoud, S. S.; Sigel, H. *Eur. J. Biochem.* **1989**, *179*, 451–458.

90. Massoud, S. S.; Liang, G.; Tribolet, R.; Sigel, H. *Rec. Trav. Chim. Pays-Bas* **1987**, *106*, 207.

91. Berger, N. A.; Eichhorn, G. L. *Biochemistry* **1971**, *10*, 1847–1857.

92. Eichhorn, G. L.; Berger, N. A.; Butzow, J. J.; Clark, P.; Rifkind, J. M.; Shin, Y. A.; Tarien, E. *Adv. Chem. Ser.* **1971**, *100*, 135–154.

93. Geraldes, C. F. G. C.; Santos, H.; Xavier, A. V. *Can. J. Chem.* **1982**, *60*, 2976–2983.

94. Sigel, H.; Martin, R. B.; Tribolet, R.; Häring, U. K.; Malini-Balakrishnan, R. *Eur. J. Biochem.* **1985**, *152*, 187–193.

95. Rogers, N. K.; Moore, G. R.; Sternberg, M. J. E. *J. Mol. Biol.* **1985**, *182*, 613–616.

96. Moore, G. R. *FEBS Lett.* **1983**, *161*, 171–175.

97. Rees, D. C. *J. Mol. Biol.* **1980**, *141*, 323–326.

98. Raudaschl-Sieber, G.; Schöllhorn, H.; Thewalt, U.; Lippert, B. *J. Am. Chem. Soc.* **1985**, *107*, 3591–3595.
99. Sheldrick, W. S.; Günther, B. *Inorg. Chim. Acta* **1988**, *152*, 223–226.
100. Sigel, H.; Fischer, B. E.; Prijs, B. *J. Am. Chem. Soc.* **1977**, *99*, 4489–4496.
101. Sigel, H. *Inorg. Chem.* **1980**, *19*, 1411–1413.
102. Fischer, B. E.; Sigel, H. *J. Am. Chem. Soc.* **1980**, *102*, 2998–3008.
103. Sigel, H. *Pure & Appl. Chem.* **1989**, *61*, 923–932.
104. Lippard, S. J. *Acc. Chem. Res.* **1978**, *11*, 211–217.
105. Barton, J. K.; Lippard, S. J. In "Metal Ions in Biology"; Spiro, T. G.; Ed.; Wiley-Interscience: New York, 1980; Vol. 1, pp 31–113.
106. Barton, J. K. *Comments Inorg. Chem.* **1985**, *3*, 321–348.
107. Yamauchi, O.; Odani, A.; Shimata, R.; Kosaka, Y. *Inorg. Chem.* **1986**, *25*, 3337–3339.
108. Mitchell, P. R.; Sigel, H. *J. Am. Chem. Soc.* **1978**, *100*, 1564–1570.
109. Sigel, H. In "Frontiers in Bioinorganic Chemistry"; Xavier, A. V., Ed.; VCH Verlagsgesellschaft: Weinheim, FRG, 1986; pp 84–93.
110. Malini-Balakrishnan, R.; Scheller, K. H.; Häring, U. K.; Tribolet, R.; Sigel, H. *Inorg. Chem.* **1985**, *24*, 2067–2076.
111. Sigel, H.; Malini-Balakrishnan, R.; Häring, U. K. *J. Am. Chem. Soc.* **1985**, *107*, 5137–5148.
112. Liang, G.; Tribolet, R.; Sigel, H. *Inorg. Chem.* **1988**, *27*, 2877–2887.
113. (a) Scheller-Krattiger, V.; Sigel, H. *Inorg. Chem.* **1986**, *25*, 2628–2634. (b) Sigel, H. *Coord. Chem. Rev.* **1990**, *100*, in press.
114. Sherman, S. E.; Lippard, S. J. *Chem. Rev.* **1987**, *87*, 1153–1181.
115. Lippard, S. J. *Pure & Appl. Chem.* **1987**, *59*, 731–742.
116. Reedijk, J.; Fichtinger-Schepman, A. M. J.; van Oosterom, A. T.; van de Putte, P. *Structure and Bonding* **1987**, *67*, 53–89.
117. Reedijk, J. *Pure & Appl. Chem.* **1987**, *59*, 181–192.
118. Sigel, H.; Erlenmeyer, H. *Helv. Chim. Acta* **1966**, *49*, 1266–1274.
119. Sigel, H.; Prijs, B.; Erlenmeyer, H. *Experientia* **1967**, *23*, 170–177.
120. Hiai, S. *J. Mol. Biol.* **1965**, *11*, 672–691.
121. Sayenko, G. N.; Babii, A. P.; Bagaveyev, I. A. *Biofizika* **1986**, *31*, 412–416; *Engl. Trans. Biophysics* **1986**, *31*, 451–456.
122. Sigel, H.; Müller, U. *Helv. Chim. Acta* **1966**, *49*, 671–681.
123. Sigel, H.; Flierl, C.; Griesser, R. *J. Am. Chem. Soc.* **1969**, *91*, 1061–1064.
124. Griesser, R.; Prijs, B.; Sigel, H. *J. Am. Chem. Soc.* **1969**, *91*, 7758–7760.
125. Erlenmeyer, H.; Müller, U.; Sigel, H. *Helv. Chim. Acta* **1966**, *49*, 681–690.
126. Erlenmeyer, H.; Griesser, R.; Prijs, B.; Sigel, H. *Biochim. Biophys. Acta* **1968**, *157*, 637–640.
127. Sigel, H. *Helv. Chim. Acta* **1967**, *50*, 582–588.

128. Sigel, H. *Angew. Chem.* **1969**, *81*, 161–171; *Angew. Chem. Int. Ed. Engl.* **1969**, *8*, 167–177.

129. Văsák, M.; Nagayama, K.; Wüthrich, K.; Mertens, M. L.; Kägi, J. H. R. *Biochemistry* **1979**, *18*, 5050–5055.

130. Fry, D. C.; Byler, D. M.; Susi, H.; Brown, E. M.; Kuby, S. A.; Mildvan, A. S. *Biochemistry* **1988**, *27*, 3588–3598.

131. Hélène, C.; Lancelot, G. *Prog. Biophys. Mol. Biol.* **1982**, *39*, 1–68.

132. Sigel, H.; Ed. "Interrelations Among Metal Ions, Enzymes, and Gene Expression"; Vol. 25 of "Metal Ions in Biological Systems"; Dekker: New York, 1989.

133. Lacey, J. C., Jr.; Mullins, D. W., Jr. *Origins of Life* **1983**, *13*, 3–42.

134. Casas-Finet, J. R.; Jhon, N.-I.; Maki, A. H. *Biochemistry* **1988**, *27*, 1172–1178.

135. Lacey, J. C., Jr.; Mullins, D. W., Jr.; Watkins, C. L. *J. Biomolecular Struct. Dynamics* **1986**, *3*, 783–793.

136. Scheraga, H. A. *Acc. Chem. Res.* **1979**, *12*, 7–14.

RECEIVED May 11, 1989

INDEXES

Author Index

Affiliation Index

Subject Index

Production: Donna Lucas
Indexing: Deborah H. Steiner
Acquisition: Cheryl Shanks

Text was typeset on an Apple Laserwriter IINT,
using a Macintosh II computer and Microsoft Word 3.0 software
Typeface: Palatino

Elements typeset by Hot Type Ltd., Washington, DC
Printed and bound by Maple Press, York, PA

Other ACS Books

Biotechnology and Materials Science: Chemistry for the Future
Edited by Mary L. Good
160 pp; clothbound, ISBN 0–8412–1472–7; paperback, ISBN 0–8412–1473–5

Polymeric Materials: Chemistry for the Future
By Joseph Alper and Gordon L. Nelson
110 pp; clothbound, ISBN 0–8412–1622–3; paperback, ISBN 0–8412–1613–4

Chemical Structure Software for Personal Computers
Edited by Daniel E. Meyer, Wendy A. Warr, and Richard A. Love
ACS Professional Reference Book; 107 pp;
clothbound, ISBN 0–8412–1538–3; paperback, ISBN 0–8412–1539–1

Personal Computers for Scientists: A Byte at a Time
By Glenn I. Ouchi
276 pp; clothbound, ISBN 0–8412–1000–4; paperback, ISBN 0–8412–1001–2

The Language of Biotechnology: A Dictionary of Terms
By John M. Walker and Michael Cox
ACS Professional Reference Book; 256 pp;
clothbound, ISBN 0–8412–1489–1; paperback, ISBN 0–8412–1490–5

Cancer: The Outlaw Cell, Second Edition
Edited by Richard E. LaFond
274 pp; clothbound, ISBN 0–8412–1419–0; paperback, ISBN 0–8412–1420–4

Practical Statistics for the Physical Sciences
By Larry L. Havlicek
ACS Professional Reference Book; 198 pp; clothbound; ISBN 0–8412–1453–0

The Basics of Technical Communicating
By B. Edward Cain
ACS Professional Reference Book; 198 pp;
clothbound, ISBN 0–8412–1451–4; paperback, ISBN 0–8412–1452–2

The ACS Style Guide: A Manual for Authors and Editors
Edited by Janet S. Dodd
264 pp; clothbound, ISBN 0–8412–0917–0; paperback, ISBN 0–8412–0943–X

Chemistry and Crime: From Sherlock Holmes to Today's Courtroom
Edited by Samuel M. Gerber
135 pp; clothbound, ISBN 0–8412–0784–4; paperback, ISBN 0–8412–0785–2

For further information and a free catalog of ACS books, contact:
American Chemical Society
Distribution Office, Department 225
1155 16th Street, NW, Washington, DC 20036
Telephone 800–227–5558